Support for the Fleet

U.S. Navy and Royal Australian Navy
Service Force Ships That
Served in Vietnam
1965–1973

During the Vietnam War, 136 U.S. Navy and three Royal Australian Navy Service Force ships served in Vietnam. It was not glamorous duty, and the men who toiled aboard the ships received little recognition. It wouldn't make good reading in Des Moines that the warships on the gunline could not fire their guns, or that carriers on Yankee and Dixie stations could not fly airstrikes against the enemy. Were it not for the sweat, heat, fatigue, and boredom endured by sailors serving in mostly old ships from World War II, that would have been the headline. These ships delivered food, fuel, ammunition, and critical supplies to the destroyers on the gunline, riverine craft patrolling inland waterways, and aircraft carriers, as well as ferrying troops in and out of the war zone, and those needing medical attention to the care of Navy Nurses on hospital ships. These brave men and their ships were often targeted by the Viet Cong specifically because they enabled the Allied forces to hold off the enemy and defend the freedom of the South Vietnamese people. In every war and military engagement, the front lines depend on replenishment. These are the people responsible for maintaining those in harm's way, putting themselves in danger to do so. This book, a companion to *On the Gunline* and *Gators Offshore and Upriver*, highlights the herculean efforts of the Service Force, whose vital contributions "on the line" have been largely overlooked by historians. Two hundred fifty-two photographs; maps and diagrams; appendices; a bibliography; and an index to full-names, places and subjects add value to this work.

Support for the Fleet

U.S. Navy and Royal Australian Navy
Service Force Ships That
Served in Vietnam
1965–1973

Cdr. David D. Bruhn, USN (Retired)

HERITAGE BOOKS
2020

HERITAGE BOOKS
AN IMPRINT OF HERITAGE BOOKS, INC.

Books, CDs, and more—Worldwide

For our listing of thousands of titles see our website
at
www.HeritageBooks.com

Published 2020 by
HERITAGE BOOKS, INC.
Publishing Division
5810 Ruatan Street
Berwyn Heights, Md. 20740

Copyright © 2020 Cdr. David D. Bruhn, USN (Retired)

All rights reserved. No part of this book may be reproduced or transmitted in any form or by any means, electronic or mechanical, including photocopying, recording or by any information storage and retrieval system without written permission from the author, except for the inclusion of brief quotations in a review.

International Standard Book Number
Paperbound: 978-0-7884-5951-1

To the officers and men of the U.S. Navy and Royal Australian Navy who served aboard Service Force ships in coastal waters off Vietnam, the Tonkin Gulf and South China Sea, and (some) in inland waters; and to the "angels of mercy" (Navy nurses) aboard the hospital ships *Repose* and *Sanctuary* offshore who, with other medical staff, provided emergency treatment and care for thousands of wounded Soldiers, Sailors, Airmen, Marines, and Coast Guardsmen

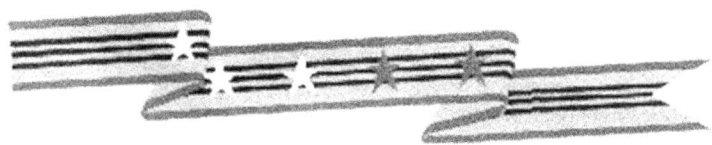

CAMPAIGNS
1. Vietnam Advisory Campaign (15 March 1962-7 March 1965)
2. Vietnam Defense Campaign (8 March-24 December 1965)
3. Vietnamese Counteroffensive (25 December 1965-30 June 1966)
4. Vietnamese Counteroffensive (1 July 1966-31 May 1967) Phase II
5. Vietnamese Counteroffensive (1 June 1967-29 January 1968) Phase III
6. Tet Counteroffensive (30 January-1 April 1968)
7. Vietnamese Counteroffensive (2 April-30 June 1968) Phase IV
8. Vietnamese Counteroffensive (1 July-1 November 1968) Phase V
9. Vietnamese Counteroffensive (2 November 1968-22 February 1969) Phase VI
10. Tet 69/Counteroffensive (23 February-8 June 1969)
11. Vietnam Summer-Fall 1969 (9 June-31 October 1969)
12. Vietnam Winter-Spring 1970 (1 November 1969-30 April 1970)
13. Sanctuary Counteroffensive (1 May-30 June 1970)
14. Vietnamese Counteroffensive (1 July 1970-30 June 1971) Phase VII
15. Consolidation I (1 Jul 1971-30 Nov 1971)
16. Consolidation II (1 Dec 1971-29 Mar 1972)
17. Vietnam Ceasefire Campaign (30 Mar 1972-28 Jan 1973)

Contents

Foreword by Dr. Salvatore R. Mercogliano	xiii
Foreword by Commodore Hector Donohue, AM RAN (Retired)	xv
Acknowledgements	xix
Preface	xxi
1. Ambush on the Co Chien River	1
2. Prelude to the Vietnam War	9
3. A Number of Firsts in 1965	17
4. The "Vung Tau Ferry"	23
5. HMAS *Boonaroo* and *Jeparit*	29
6. Survey Ship Operations	37
7. USS *Maury* and *Serrano* Return to Vietnam	51
8. Final Survey Work in Vietnam	59
9. Seaplane Surveillance Operations	63
10. AGTRs *Oxford* and *Jamestown*	73
11. Major Communications Relay Ships	79
12. *Patapsco*-class Gasoline Tankers	91
13. Stores, Combat Stores, and General Stores-Issue Ships	107
14. Early War Salvage Efforts	119
15. Royal Australian Navy Clearance Diving Team 3	133
16. Hospital ships *Repose* and *Sanctuary*	143
17. Mobile Shipyards	157
18. Ammunition Ships	171
19. Fleet Oilers	191
20. Fast Combat Support Ships and Replenishment Oilers	207
21. War's End	219
Postscript (66 photographs of shipboard/liberty activities)	225
Appendices	
A. Justification for Unit Awards Received by USS *Cohoes*	265
B. U.S. Seventh Fleet Service Force Ships' Unit Awards	267
C. Harbor Clearance Unit One Meritorious Unit Awards	271
D. ABCD Anthony L. Ey, RAN Letter of Commendation	273
E. USS *Sanctuary* Meritorious Unit Commendation	275
F. USS *Ajax* and USS *Jason* Navy Unit Commendation	277
G. USS *Chara* Meritorious Unit Commendation	279
H. USS *Mount Katmai* Meritorious Unit Commendation	281
I. U.S. Seventh Fleet Service Force Personnel Casualties	283
Bibliography/Notes	285
Index	303

About the Author   317

Photos and Illustrations

Preface-1: Commander, Service Force U.S. Pacific Fleet plaque   xxi
Preface-2: *On the Gunline* & *Gators Offshore and Upriver* book covers   xxii
Preface-3: Three-ship underway replenishment group   xxiv
Preface-4: USS *Sacramento* transferring fuel and ammo to *Hancock*   xxv
Preface-5: Gasoline tanker USS *Elkhorn* under way   xxviii
Preface-6: Net laying ship USS *Cohoes* off Hawaii   xxx
Preface-7: Salvage ship USS *Bolster* with a diver in the water   xxxii
Preface-8: *Bolster*'s dive boat towing a fuel line into position   xxxiii
Preface-9: Rear Admirals Edwin Hooper and Norvell Ward   xxxiii
Preface-10: Pacific Ocean crossing aboard USS *Mount Baker*   xxxiv
1-1: Painting by Richard DeRosset of USS *Brule* under attack   1
1-2: USS *Brule* at anchor in South Vietnamese waters   2
1-3: USS *Garrett County* on the Co Chien River   3
1-4: USS *Brule* on the Bassac River   6
1-5: Browning .50-caliber heavy machine gun   7
1-6: Crewmembers of the USS *Mark* offloading supplies   8
2-1: Poster of Ho Chi Minh, ruler of North Vietnam   9
2-2: Boat landing on the Mekong River, French Indochina   10
2-3: French Adm. Thierry D'Angenlieu and Gen. Richard Brunot   12
2-4: Refugees board the tank landing ship USS *LST-516*   14
2-5: General and flag officers await arrival of refugees in Saigon   15
2-6: General Westmoreland and Admirals Moorer and Veth   16
3-1: Camp Holloway, Republic of Vietnam   18
3-2: A-4 Skyhawk preparing to launch from USS *Coral Sea*   18
3-3: U.S. Marines wade ashore at Da Nang, South Vietnam   19
3-4: Painting *Operation Market Time* by Gene Klebe, 1965   20
3-5: Paratroopers of the 173rd Airborne Brigade   21
3-6: Sky Soldiers aboard the USNS *General William A. Mann*   21
3-7: Destroyer USS *Henry W. Tucker* under way   22
4-1: Light fleet aircraft carrier HMAS *Sydney*, circa 1950-1953   23
4-2: 1 RAR formed up on the flight deck of HMAS *Sydney*   25
4-3: DeLong Pier at Vung Tau with HMAS *Jeparit* alongside   26
4-4: HMAS *Sydney* in her role as the "Vung Tau Ferry"   27
4-5: Swimmer-sappers captured in the Vung Tau port area   28
5-1: Australian motor vessel *Boonaroo*   29
5-2: HMAS *Jeparit* under way in Sydney Harbour, Australia   34
5-3: *Jeparit* at Vung Tau, South Vietnam   34
5-4: German raider HK *Pinguin* in the Indian Ocean   36

6-1: USS *Maury* at anchor, with *Serrano* alongside	37
6-2: *Maury* conducts tasks preparatory to survey operations	39
6-3: USS *Maury* under way, location and date unknown	40
6-4: Five ocean minesweepers of Mine Division 73	41
6-5: Memorial containing a splinter from Magellan's cross	42
6-6: Cam Ranh Bay, Vietnam, November 1965	43
6-7: RAYDIST transmitting station on Hon Ngoi Island	43
6-8: Vietnamese command junk on patrol	44
6-9: Members of the Vietnamese "junk force"	45
6-10: Chu Lai Peninsula, Vietnam, January 1966	46
6-11: Machine gunner and lookout/loader aboard Soundboat *7*	47
6-12: Hong Kong and Kowloon by night	49
7-1: USS *Serrano* moored along larger running mate *Maury*	52
7-2: Survey ship USS *Maury*'s Soundboat *8*	53
7-3: USCG cutter *Point Kennedy*	54
7-4: UH-1B Iroquois helicopters aboard USS *Harnett County*	55
7-5: Beachside bars in Vung Tau, South Vietnam	55
7-6: Main street in Vung Tau	56
7-7: Nha Trang and some *Maury* crewmembers ashore	56
7-8: Yokosuka nightlife, and liberty by day in Yokosuka	58
8-1: Survey ship USS *Tanner*, with *Sheldrake* alongside	59
8-2: Rear Adm. Norvell G. Ward aboard USS *Tanner*	61
8-3: American actress, model, and sex symbol Mamie Van Doren	62
9-1: A Marlin seaplane lifts off Cam Ranh Bay, South Vietnam	63
9-2: USS *Pine Island* hoisting aboard a seaplane for maintenance	64
9-3: Rear Adm. Roy Maurice Isaman, USN	65
9-4: USS *Currituck* off Con Son Island in 1965	67
9-5: SP-5B Marlin seaplane on the water in Cam Ranh Bay	67
9-6: SP-5B Marlin alongside seaplane tender USS *Pine Island*	68
9-7: Marker where General Buckner was killed on Okinawa	71
9-8: A P-3 Orion flying a routine patrol off Cam Ranh Bay	72
10-1: Technical research ship USS *Oxford* under way	73
10-2: Naval Air Station, Sangley Point, Philippines	74
11-1: USS *Annapolis* under way in New York Harbor	79
11-2: USS *Arlington* under way, circa 1967	81
11-3: USS *Arlington* as viewed by astronauts from recovery ship	83
11-4: Descent of the Apollo 10 space capsule	84
11-5: U.S. Naval Air Station, Midway	86
11-6: President Richard M. Nixon and entourage at Midway Island	86
11-7: President Nixon and President Thieu shaking hands	87
11-8: A Navy helicopter picks up the Apollo 11 astronauts	88
11-9: Apollo 11 astronauts enter Mobile Quarantine Facility	89

12-1: Gasoline tanker USS *Noxubee* under way 91
12-2: Pearl Harbor, with Diamond Head in the background 93
12-3: USS *Kishwaukee* taking fuel from a USNS tanker 95
12-4: Fuel storage area, partially enclosed by stacked bags of sand 96
12-5: U.S. Army barge at Sa Huynh, South Vietnam 96
12-6: Fishing village of Sa Huynh 97
12-7: Fuel farm at Cua Viet under enemy artillery fire 97
12-8: Subic Bay, on the west coast of the island of Luzon 98
12-9: Spanish Gate, U.S. Naval Base, Subic Bay, Philippines 98
12-10: Da Nang Harbor, South Vietnam, in January 1966 99
12-11: *Kishwaukee* returning to Pearl Harbor from deployment 100
12-12: Naval Support Activity, Da Nang, Cua Viet detachment 101
12-13: USS *Patapsco* at anchor off Cua Viet, South Vietnam 106
13-1: Romeo flags closed up on a replenishment ship 107
13-2: USS *Castor* with USS *Pollux* berthed astern 109
13-3: USS *Castor* at berth six, in India Basin, at Sasebo, Japan 110
13-4: Boatswain's Mates at opening of *Castor*'s No. 2 Hold 111
13-5: Cargo being staged for transfer to a ship alongside 112
13-6: Transfer of materials via span wire and cargo net 113
13-7: Stores ship USS *Pictor* berthed in port 114
13-8: *Pictor*'s helmsman during an underway replenishment 115
13-9: Landing craft approaching *Pictor* for food transfer 115
13-10: Landing craft alongside the stores ship *Pictor* 116
13-11: USS *Niagara Falls*, from pieces of steel to a proud ship 117
13-12: Banner on bridge wing of *Niagara Falls* 118
14-1: Destroyer USS *Frank Knox* aground on Pratas Reef 120
14-2: *Frank Knox* aground with a helicopter overhead 121
14-3: Tank landing ship USS *Terrell County* under way 126
14-4: Fleet auxiliary tug USS *Mahopac* under way 127
14-5: Fleet ocean tug USS *Hitchiti* under way 128
14-6: Tank landing ship USS *Coconino County* 130
14-7: Transport USNS *Geiger* arriving at Rota, Spain 131
14-8: USS *Abnaki* keeping a Soviet trawler under surveillance 132
15-1: Painting *Ambush on the Long Tau* by Richard DeRosset 133
15-2: Minesweeping boat *MSB-49* on the Saigon River 136
15-3: Australian Army tank being unloaded from *YFU-63* 138
15-4: RAN Clearance Diving Team 3 (8th Contingent) 139
15-5: Members of CDT 3 attaching a jackstay to an Army craft 140
15-6: Vung Tau Harbor Entrance Control Post 141
15-7: CDT 3 8th Contingent arriving at Mascot Airport, Sydney 142
16-1: Hospital ship USS *Sanctuary* at Mare Island Naval Shipyard 143
16-2: USS *Sanctuary* anchored at Wakayama Harbor, Japan 144

16-3: HMS *Amethyst* leaving Chinnampo, North Korea   145
16-4: Hospital ship USS *Repose* anchored in Da Nang Harbor   147
16-5: Casualties airlifted to *Repose* by helicopter for treatment   148
16-6: Vice Adm. Robert B. Brown, MC, USN   150
16-7: Navy Captains John F. Collingwood and Gerald J. Duffner   151
16-8: Navy Captains Robert F. Menge and Arthur J. Draper   152
16-9: Navy nurses Mary Lukacs, Carol Crandal, and Janice Quinn   152
16-10: Patients being transported off hospital ship *Sanctuary*   154
16-11: Operating room aboard hospital ship USS *Repose*   155
16-12: Crew boarding USS *Sanctuary* during her recommissioning   156
17-1: Repair ship USS *Jason* lying at anchor off Vung Tau   157
17-2: Chu Lai, South Vietnam, August 1967   159
17-3: Repair ship USS *Markab* under way   160
17-4: USS *Markab* sailors at a vendor's stand in Vung Tau   163
17-5: Popular beach at Vung Tau   164
17-6: A UH-1 Iroquois ("Seawolf") helicopter in flight   167
17-7: Repair ship USS *Hector* at Vung Tau, South Vietnam   169
17-8: Destroyer tenders USS *Samuel Gompers* and USS *Piedmont*   170
18-1: USS *Mount Hood* explodes in Seeadler Harbor   171
18-2: Ammunition ship USS *Mount Baker* under way   172
18-3: Ammunition ship USS *Mazama* alongside an aircraft carrier   174
18-4: Welcome home of *Mazama* from deployment to Vietnam   175
18-5: USS *Virgo* under way in Subic Bay, Philippines   175
18-6: Admiral Zumwalt, USN, and Rear Admiral Salzer, USN   177
18-7: Highline transfer of bombs and rockets between ships   179
18-8: USS *Vesuvius* with San Francisco Golden Gate Bridge astern   180
18-9: Ammunition ship USS *Suribachi* under way   182
18-10: Battleship USS *New Jersey* receiving powder for her guns   183
18-11: USS *Kilauea* off Quonset Point, Rhode Island   183
18-12: Comdr. William L. McGonagle, USN, aboard USS *Liberty*   184
18-13: Naval Magazine, Subic Bay, on Camayan Point   186
18-14: POWs Col. Robinson Risner and Capt. James Stockdale   189
19-1: Fleet oiler USS *Mattaponi* near San Diego, California   191
19-2: *Mattaponi* sailors involved in an underway replenishment   192
19-3: Fleet oiler USS *Kennebec* at sea   193
19-4: USS *Tappahannock* transferring jet fuel to USS *Enterprise*   194
19-5: "Jumboized" USS *Ashtabula* under way off San Diego   195
19-6: Bow being joined to the fleet oiler USS *Caloosahatchee*   196
19-7: Fleet oiler USS *Kawishiwi* under way off Oahu, Hawaii   198
19-8: Fleet oiler USS *Ponchatoula* under way off Oahu   199
19-9: Entering Subic Bay, and mooring at the naval station   201
19-10: "Philatelic mail" commemorating the Apollo 8 flight   202

19-11: USS *Ashtabula* encountering Typhoon TESS   204
19-12: Fleet tug USS *Cree* blocking a Soviet trawler's movements   205
19-13: Barbara Eden in the 1960s, and with Bob Hope in 1987   206
20-1: USS *Hancock* and *Robison* refuel from the USS *Sacramento*   207
20-2: POL (Petroleum, oil and lubricants) Pier at Subic Bay   209
20-3: A carrier and a minesweeper refuel from *Sacramento*   210
20-4: Bearded, khaki-clad *Sacramento* crewmember   211
20-5: USS *Carpenter* and *Hancock* replenish from USS *Camden*   213
20-6: Replenishment oiler USS *Wichita* in heavy seas   214
21-1: Sailor hauling down a flag symbolizes war's end in 1973   219
21-2: Commencement of the Vietnam Peace talks in early 1969   220
21-3: Jacket patches associated with Operation END SWEEP   222

Maps and Diagrams

Preface-1: Locations of Yankee and Dixie Stations   xxvi
Preface-2: I-Corps, and part of II-Corps in South Vietnam   xxix
5-1: Australia   31
6-1: Con Dao Archipelago   46
7-1: South Vietnam, and surrounding areas   57
8-1: Western Pacific   60
10-1: (Dao) Phu Quoc Island in the Gulf of Thailand   77
12-1: Abbreviated map of South Vietnam   94
16-1: Amphibious landings in northern area of South Vietnam   153
17-1: Track chart of USS *Markab*'s 1967 WestPac deployment   161
17-2: Rung Sat Special Zone   165

Vietnam sunset.
USS *Oklahoma City* (CLG-5) Western Pacific 1971-1972 cruise book

Foreword

David Bruhn's *Support for the Fleet: U.S. and Royal Australian Navy Service Force Ships That Served in Vietnam, 1965 – 1973* follows his two previous works that detail vital aspects of the naval side of the Vietnam War. In *Gators Offshore and Upriver* and *On the Gunline* he examined the important role played by amphibious warfare and naval close support throughout the conflict. In *Support for the Fleet* he takes us from Subic Bay in the Philippines to Yankee and Dixie Stations off the coast of Vietnam, to the inland water ways of Southeast Asia to document the unheralded, yet essential role that logistics and naval auxiliaries played in prosecuting the war in Vietnam by both the United States and Royal Australian navies. He examines all aspects of this type of warfare, from surveillance operations, to fleet support, and salvage operations.

The naval aspect of the Vietnam War has not garnered comprehensive attention by historians or the general public throughout the years. The quintessential images of the naval war are aircraft catapulting from carriers assigned to Task Force 77 or river patrol boats (PBRs) streaking up a river. Yet, the role played by naval auxiliaries in the Vietnam War were essential to the conduct of the fighting and exposed these vessels to great dangers in the execution of their missions. Not since the Civil War had the U.S. Navy waged a coastal and riverine campaign, with a large offshore presence to the scale conducted in Vietnam. David Bruhn captures this from the outset with an attack on USS *Brule* (AKL-28) on the Co Chien River. *Brule* was an ex-Army freight-supply ship, best depicted in the movie *Mr. Roberts* as USS Reluctant – banished to supply the backwater ports of the Pacific. Unlike Mr. Roberts, action found the crew of *Brule* on the morning of August 24, 1968 when she came under rocket fire. She sustained damage, but no casualties, and was able to complete her resupply mission.

The attack on USS *Brule*, along with numerous other instances, and a wealth of information and data on the ships of the U.S. Navy Service Forces, along with those of the Royal Australian Navy – including the famed Vung Tau Ferry – make *Support for the Fleet*, along with Bruhn's earlier works required reading for all those studying military operations in the Vietnam War. The contribution of the ships and crews was essential to the execution of the military strategy undertaken by the

United States during the Vietnam War. Of importance, is the understanding of the logistical undertaking required to support the deployment of such a large force to Southeast Asia over the course of eight years. The scope of the mission is indicated by the service of USS *Hassayampa* (AO-145). Over the course of several deployments during the war, *Hassayampa* completed a total of 64 combat tours (termed "swings on the line" by the Service Force) in delivering fuel to warships offshore, earning a Meritorious Unit Commendation.

Stories like *Hassayampa* and *Brule* make us appreciate the role of the sailors who crewed their naval auxiliaries during the Vietnam War. Rear Admiral Worrall Reed Carter titled his work on naval logistics during the Second World War, *Beans, Bullets, and Black Oil*. In that same vein, *Support to the Fleet* details Vietnam naval logistics down to all levels of naval auxiliary vessels and elevates their role alongside the carriers and PBRs of the Vietnam War.

Salvatore R. Mercogliano, Ph.D., Campbell University
Author, *Fourth Arm of Defense, Sealift and Maritime Logistics in the Vietnam War*

Foreword

Support for the Fleet is the third in David Bruhn's books on the naval aspects of the war in Vietnam. It rounds out the comprehensive analysis of the Navy's contribution to the war contained in his earlier books: *On the Gunline* and *Gators Offshore and Upriver*. Support tasks are never glamorous and are usually underrated by both the services and historians. The neglect of support, including logistics, as a field of serious enquiry has generally been the norm amongst military historians, and Bruhn should be commended for extending his research to include this discipline.

His final book in the trilogy covers a wide range of support activities. The majority of the materiel supporting the Vietnam War came by sea. This included most of the ammunition and fuel as well as the supplies, vehicles, and construction resources consumed by the massive allied war effort. With primary responsibility for the sea lines of communication to Southeast Asia, the U.S. Navy oversaw the development of a 7,000 nautical mile (nm) transoceanic lifeline to U.S. forces fighting ashore, steaming in the South China Sea, and to bases throughout the Pacific. The Australian support activity was more modest, but nevertheless important because the deployed Australian forces were, for the first time, fully supported from home.

The task of moving, supplying and maintaining Australian forces in Vietnam was predominantly undertaken by the Royal Australian Navy (RAN) and ships taken up from trade from the Australian National Line (ANL). In addition, the Royal Australian Air Force (RAAF) maintained a regular C-130 courier service as did Australia's international airline, Qantas Airways Ltd.

HMAS *Sydney* the former aircraft carrier was the mainstay of naval logistics support operations for Australian forces in Vietnam. *Sydney* was commissioned in 1948 and was central in the development of Australia's post-war naval aviation capability. There was considerable public interest in the planned acquisition of two aircraft carriers including speculation as to the names of the carriers, which appeared in the *Sydney Morning Herald* in December 1947, suggesting they would be named after Australian statesmen. A follow-up story suggested this had occurred; as the Navy had advised, the first one was called *Terrible*. On transfer to the RAN, HMS *Terrible* became HMAS *Sydney*.

Sydney served with distinction in the Korean War, but by 1958 with the RAN facing severe financial constraints, she was placed in reserve, a mere decade after commissioning.

Sydney recommissioned as a fast troop transport in early 1962, with her first operational deployment in mid-1964 deploying Australian Army units to Malaysia as part of Australia's initial contribution to the Indonesian Confrontation. *Sydney* began her first voyage to Vietnam in May 1965, transporting 1st Battalion Royal Australian Regiment (RAR) from Sydney to Vung Tau.

Sydney's route was north through the Coral Sea then west passing south of the Philippines and north of Indonesia, a voyage of some 5,000 nm. From the Coral Sea, she sailed west through the Vitiaz Strait between New Britain and Papua New Guinea's north eastern Huon Peninsula, south of Basilan Island in the southern Philippines, then through Balabac Strait between Sabah and Palawan Island in the Philippines, and across the South China Sea to Cap St. Jacques and the port of Vung Tau. Later voyages included stops at Manus Island. RAR Battalions were embarked in Sydney, Brisbane, Townsville and Adelaide. For the Adelaide embarkation, *Sydney* sailed south about Australia and north through the Indian Ocean, through the Sunda Strait in Indonesia, transiting east of Malaysia and north to Vung Tau. From June 1969 until her final voyage, *Sydney* made her northern passage inside the Great Barrier Reef, west across northern Australia and north through the Sunda Strait and on to Vietnam.

By 1966 with Australian ground forces well established in Vietnam, *Sydney* began a regular pattern of disembarking one battalion at Vung Tau and back loading another for the return passage to Australia. In the early days *Sydney*'s turnaround in Vung Tau took two days, but this was gradually reduced until, by 1967, the unloading and back loading of men and equipment generally took only half a day. She was quickly known and remembered fondly by those involved as the Vung Tau Ferry. Australia's combat role in South Vietnam ceased in March 1972 when *Sydney* brought home the last combat elements. *Sydney* returned to Vung Tau for one final visit in November 1972, when she delivered a cargo of defence aid for Vietnam and Cambodia. Between 1965 and 1972 *Sydney* undertook 24 voyages to Vietnam, transporting 16,094 troops some 6,000 tonnes of cargo and 2,375 vehicles during this period. She was decommissioned in November 1973.

The transporting of equipment and stores in *Sydney* was supplemented by chartering two ANL cargo ships, MV *Boonaroo* and MV *Jeparit* and through the deployment of three Army landing ships medium: *Vernon Sturdee*, *Harry Chauvel* and *Clive Steele*, and the Australian Army cargo ship *John Monash*. The Army vessels were operated by 32nd Small Ships Squadron, Royal Australian Engineers.

MV *Boonaroo* completed one round-trip voyage to Vietnam in mid-1966. She was commissioned as HMAS *Boonaroo* in March 1967 after members of the Seaman's Union refused to sail the ship to Vietnam with a cargo of RAAF ordnance onboard. Following that trip, she was handed back to the ANL.

MV *Jeparit* commenced transporting equipment and supplies in mid-1966. After five voyages, some merchant seamen refused to man the vessel. To resolve this issue, crew members who were prepared to continue to serve in *Jeparit*, were supplemented by a RAN detachment. Following *Jeparit*'s 26th round voyage to Vietnam, further industrial trouble developed and the ship was commissioned as HMAS *Jeparit* in December 1969. She continued to operate with a mixed merchant navy/RAN crew until she was handed back to the ANL in March 1972, having completed a total of 38 trips to Vietnam.

The significant efforts of *Sydney* and *Jeparit* made the Australian contribution to the war possible because they transported infantry battalions, supporting units, stores and equipment to and from Vietnam. The other units involved were an essential adjunct to the primary logistics support activity, and their role was vital to the successful maintenance of the Australian forces in Vietnam.

The logistic support function provided by Australia for its forces during the Vietnam War was important for two reasons. Firstly, it maintained the Australian air and ground commitment in the field and perhaps more importantly for the long term, it demonstrated for the first time that Australia was capable of transporting, maintaining and reinforcing a significant ground and air force at a distance, for a prolonged period of time with and from its own resources.

Commodore Hector Donohue AM RAN (Rtd)

Acknowledgements

Masterful maritime and aviation artist Richard DeRosset created the cover art for *Support for the Fleet*. His brilliant painting depicts the light cargo ship USS *Brule* engaged in combat with Viet Cong forces on both banks of the Co Chien River near Vinh Long. I am grateful to him for his fine work, and to CWO2 Thomas L. Young, USN (Retired), who provided details about the action. As the ship's first lieutenant/cargo officer/gunner officer, Young directed the actions of her gun crews, which silenced the rocket and automatic-weapons fire erupting from concealed positions. *Brule* suffered material damage, but no personnel casualties, owing to the rapid counterfire.

Many thanks to Dr. Salvatore R. Mercogliano, an associate professor of history at Campbell University in North Carolina and adjunct professor at the U.S. Merchant Marine Academy, for his review of the manuscript, and the perspective he offers in his foreword. Author of the book, *Fourth Arm of Defense: Sealift and Maritime Logistics in the Vietnam War*, and many articles in professional journals, Mercogliano's interest and expertise in maritime and naval subjects began when he was a deck officer aboard merchant vessels following graduation from the State University of New York Maritime College.

Commodore Hector Donohue, AM RAN (Retired), provides valuable senior Royal Australian Naval officer perspective and assessment in his foreword. The author of several books and articles, he graciously contributed much material to *Support for the Fleet*—as he earlier had for *Nightraiders*, *On the Gunline*, and *Gators Offshore and Upriver*. Donohue began his career in the RAN in 1955 as a seaman officer and subsequently sub-specialized as a clearance diver and torpedo and anti-submarine officer. His service in the RAN included command of the destroyer escort HMAS *Yarra* and the guided missile frigate HMAS *Darwin*. Ashore, he held a number of senior positions in Defence policy and force development prior to retirement in mid-1991.

I am also appreciative of assistance provided by Donald (Scotty) Allan, a former RAN clearance diver and current historian for the Royal Australian Navy Clearance Divers Association. Scotty has recently authored, with John Kennett, *ALL IN THE LINE OF DUTY: Honours and Awards to the Royal Navy's Clearance Diving Branch 1951-2018*.

Commander Lee Foley, USN (Retired), author of the recently published book *Mustang*, reviewed the manuscript with the critical eye of someone with much sea duty. After serving aboard ten ships while progressing from Seaman Recruit to Master Chief Petty Officer, Warrant Officer WO1 to CWO4, and Limited Duty Officer LTJG to LCDR, he was selected for conversion to unrestricted line and command of the ocean minesweeper USS *Excel* (MSO-439). Following that tour, he served as executive officer aboard the USS *Kansas City*, near the end of a thirty-three-year career in the Navy.

Several individuals also assisted with the book. Capt. Philip M. Palmer, USN (Retired), provided a firsthand account of his experiences while serving as commanding officer, U.S. Naval Magazine, Subic Bay.

Albert Moore has lent much help over the years, and did so again for this book. The founder and former long-serving president of the Mobile Riverine Force Association, Moore served in several units in Vietnam, including duty aboard the USS *Benewah* (APB-35), the flagship of commander, Mobile Riverine Force.

Retired Navy Photographer's Mate First Roland Nino Martinez provided information related to the Movement Report Office (a part of the U.S. Naval Forces Philippines command), which monitored all shipping, as well as surveillance of the movement of the Chinese government as they began occupying a majority of the atolls in the South China Sea as early as 1956.

Lew Smith and Jim Dresser allowed use of their photographs of "sky soldiers" of the 173rd Airborne Brigade, the first U.S. combat troops to arrive in Vietnam by ship and transport aircraft.

Finally, much thanks to my editor Lynn Marie Tosello. Her keen eye, piercing intellect, active pen, and appreciation for eloquence and prose helps me steer a straight course. She also provides much value by identifying items not adequately explained to readers who have not had the pleasure of trodding the deck of a Navy ship.

Preface

What is so incredible about the US Fleet is that it can stay somewhere forever and resupply by ships and never see land and never have to go to land, except for the supply ships resupplying. They have to have a base like Subic [Bay, Philippines] to get their supplies or else have other ships come out and resupply them. It is a dangerous operation but we do it. We learned the tricks. It is very hazardous. Fortunately, no one has ever been blown up or anything.

—Capt. Leon Grabowsky, USN (Retired), former commander, Service Squadron Five, remarking on the resupply of 7th Fleet ships in the Tonkin Gulf during the Vietnam War.[1]

Photo Preface-1

Commander, Service Force U.S. Pacific Fleet plaque.
Naval History and Heritage Command photograph #NH 73860

Support for the Fleet is the final book in a trilogy devoted to the U.S. and Royal Australian Navy ships that served in Vietnam. The first two books, *On the Gunline* and *Gators Offshore and Upriver*, detail the activities of the gunfire support ships off Vietnam, and the amphibious ships that landed Marines along its coast. Some of the "gators," primarily World War II vintage, shallow-draft tank landing ships, also plied dangerous inshore waters. Their duties including hauling supplies to far-flung bases and units in the Mekong Delta, and providing direct support for river patrol boats and "gunships" (assault helicopters).

Photo Preface-2

On the Gunline, and *Gators Offshore and Upriver* book covers; created by Heritage Books designer/editor Debbie Riley, based on paintings by maritime artist Richard DeRosset.

During the Vietnam War, 270 U.S. Navy and four Royal Australian Navy warships served at various times on the gunline. Within this armada were the battleship *New Jersey*, 10 cruisers, 212 destroyers, 50 destroyer escorts, and the inshore fire support ship *Carronade*. When necessary, naval guns poured out round after round, until their barrels overheated and turned red, exterior paint blistered, and rifled-barrel liners were worn smooth. Facing and often dueling with enemy artillery, these ships collectively earned over 500 combat action ribbons.

The 142 amphibious ships that served in Vietnam garnered over 160 combat action ribbons. Many of these were earned by tank landing ships running the rivers, exposed while doing so, to ambush by recoilless-rifle, automatic-weapons, or rocket fire from the enemy hidden in jungle canopy or foliage along the banks.

THE U.S. NAVY'S SERVICE FORCE IN VIETNAM

The 135 Service Force ships to which this book is devoted, are identified in the next several pages by their general category, followed by specific function. Fleet replenishment ships made up over half of the Service Force ships that served in Vietnam, and the bulk of them were either oilers or ammunition ships, whose job it was to keep their customers steaming and shooting. Small numbers of stores ships, gasoline tankers, and multifunction ships (AOEs and AORs) comprised the rest.

FLEET REPLENISHMENT SHIPS (Seventy-two)

AE   Ammunition Ships – 24
Butte (AE-27), *Chara* (AE-31), *Diamond Head* (AE-19), *Firedrake* (AE-14), *Flint* (AE-32), *Great Sitkin* (AE-17), *Haleakala* (AE-25), *Kilauea* (AE-26), *Mauna Kea* (AE-22), *Mauna Loa* (AE-8), *Mazama* (AE-9), *Mount Baker* (AE-4), *Mount Hood* (AE-29), *Mount Katmai* (AE-16), *Nitro* (AE-23), *Paricutin* (AE-18), *Pyro* (AE-24), *Rainier* (AE-5), *Santa Barbara* (AE-28), *Shasta* (AE-6), *Suribachi* (AE-21), *Vesuvius* (AE-15), *Virgo* (AE-30), *Wrangell* (AE-12)

AF   Stores Ships - 8
Aludra (AF-55), *Bellatrix* (AF-62), *Graffias* (AF-29), *Pictor* (AF-54), *Procyon* (AF-61), *Regulus* (AF-57), *Vega* (AF-59), *Zelima* (AF-49)

AFS   Combat Stores Ships - 4
Mars (AFS-1), *Niagara Falls* (AFS-3), *San Jose* (AFS-7), *White Plains* (AFS-4)

AO   Fleet Oilers - 23
Ashtabula (AO-51), *Cacapon* (AO-52), *Caliente* (AO-53), *Chemung* (AO-30), *Chipola* (AO-63), *Cimarron* (AO-22), *Guadalupe* (AO-32), *Hassayampa* (AO-145), *Kawishiwi* (AO-146), *Kennebec* (AO-36), *Manatee* (AO-58), *Marias* (AO-57), *Mattaponi* (AO-41), *Mispillion* (AO-105), *Navasota* (AO-106), *Neches* (AO-47), *Passumpsic* (AO-107), *Platte* (AO-24), *Ponchatoula* (AO-148), *Taluga* (AO-62), *Tappahannock* (AO-43), *Tolovana* (AO-64), *Waccamaw* (AO-109)

AOE   Fast Combat Support Ships - 2
Camden (AOE-2), *Sacramento* (AOE-1)

AOG   Gasoline Tankers - 6
Elkhorn (AOG-7), *Genesee* (AOG-8), *Kishwaukee* (AOG-9), *Noxubee* (AOG-56), *Patapsco* (AOG-1), *Tombigbee* (AOG-11)

AOR   Replenishment Oilers - 5
Kansas City (AOR-3), *Milwaukee* (AOR-2), *Savannah* (AOR-4), *Wabash* (AOR-5), *Wichita* (AOR-1)

The duty of fleet replenishment ships was arduous, repetitious, and potentially dangerous, owing to cargos of flammable fuels or explosives, and repeated operations in close quarters with other ships. Duty off Vietnam for Service Force ships newly arrived in the Western Pacific, was typically preceded by "loading out" at Subic Bay, followed by the first of many "swings on the line" in a single deployment.

The expression "on the line" referred to duty in the combat zone. Ships were "off the line" when they left Vietnamese waters bound for Subic Bay or other ports for required maintenance, resupply, or crew liberty. Cruisers and destroyers enjoying a reprieve from duty "on the gunline," or gators (amphibious ships) operating offshore or upriver, might be tasked with other Seventh Fleet duties, or enjoy visits to one or more exotic Far East liberty ports. A "swing on the line" or "line swing" was unique to Service Force ships. With their services always in great demand, their offline time was normally spent at Subic loading out before returning to Vietnam, then sailing to the war zone, delivering product, sailing back to Subic to replenish it, and repeat. A swing was a single period (tour of duty) on the line.

Upon arrival in South Vietnamese waters, replenishment ships called at ports along the coast as necessary to meet demands ashore, then refueled, rearmed, or re-provisioned fleet units in the Tonkin Gulf. The replenishment of combatants at sea, via underway replenishment (UnRep) alongside or vertical replenishment (VertRep) using ship-based helicopters, enabled the ships to operate for long periods at Yankee and Dixie Stations, on MARKET TIME patrols off the coast, and on the naval gunfire support line off South Vietnam.

Photo Preface-3

USS *Pictor* (AF-54), in the foreground, highlines cargo to the USS *Camden* (AOE-2), center, while the USS *Hassayampa* (AO-145) replenishes *Camden*'s liquid cargo, 1969. National Archives photograph #XFC-02171-2-69

Photo Preface-4

USS *Sacramento* (AOE-1) transfers fuel and ammunition through hoses and highlines to USS *Hancock* (CVA-19), while a CH-46 Sea Knight helicopter supports crated Mark 77 fire bombs (napalm) being transferred to the carrier, 9 August 1966.
National Archives photograph #K-31344

YANKEE AND DIXIE STATIONS

Yankee Station was a point off Vietnam from which U.S. carrier aircraft flew strikes into North Vietnam. As shown on the map (next page), it was initially located at position 16°N, 110°E, off the coast of South Vietnam and south of the DMZ (nominally described as being at "the 17th parallel") from 1964 to 1966. However, with a massive increase in air operations over North Vietnam in 1966 associated with Operation ROLLING THUNDER, the station was moved northwest to 17°30'N, 108°30' E, about ninety miles off the North Vietnamese coast.[2]

Dixie Station was established on 15 May 1965 (at 11°N, 110°E) about eighty miles southeast of Cam Ranh Bay. In contrast to Yankee Station, from which bombing missions were flown against targets in the North, those originating from Dixie Station were in support of allied ground forces engaged in combat in South Vietnam. The strike aircraft were usually vectored to their targets by a forward-based air controller. Aircraft carriers continued to use Dixie Station in support of friendly forces until 3 August 1966, when sufficient land-based aircraft were available, and carriers were no longer needed in the area. Yankee Station remained in use until August 1973.[3]

Map Preface-1

Locations of Yankee and Dixie Stations during the Vietnam War

REPLENISHMENT PRACTICES

> *The way the Seventh Fleet staff had set it up we were bringing half the loads back to Subic instead of keeping them in the Tonkin Gulf; so I organized a system to change this to where we would have main storage ships in the Tonkin Gulf. These storage ships would keep all the assets from the ships that were going back—transfer the materials and keep them there. We needed special kinds of ships to do this.*
>
> *When the ships went out there they started at the south and came up north where they were relieved of their assets in the Gulf. The assets were then transferred to these storage ships. It involved a hell of a lot of handling and the skippers didn't like it, but it was the most efficient way to do it. We saved billions of dollars, I think, because as soon as they were through with their circuit of the coast and unloaded once in Tonkin Gulf, they could head back to Subic to reload and be ready for the next run.*
>
> —Capt. Leon Grabowsky, USN (Retired), describing the transfer of remaining cargo aboard Seventh Fleet replenishment ships to Military Sea Transportation Service ships on station in the Tonkin Gulf, before departure of the former to Subic.[4]

> *We were utterly dependent upon the sea logistical line.*
>
> —Gen. William Westmoreland (commander, U.S. Military Assistance Command, Vietnam) in *Report on Operations in South Vietnam January 1964 – June 1968.*

The supply line that provided logistics support to military forces ashore in Vietnam stretched 7,000 miles across the Pacific from the West Coast of the United States. The trans-Pacific logistics operation was carried out by merchant ships of the Military Sea Transportation Service (MSTS); renamed the Military Sealift Command in 1970. The ammunition, POL (Petroleum, oil and lubricants), and food they carried were delivered to major ports on the South China Sea (Da Nang, Qui Nhon, and Cam Ranh), and to Saigon, forty-five miles inland up the Long Tau and Saigon Rivers. Other smaller ports were also later utilized, following adequate port development.[5]

Merchant ships also delivered to Subic Bay, where Seventh Fleet replenishment ships loaded out for the resupply of naval forces inland, via delivery to South Vietnamese ports, and the fleet units off the coast.

One important interface between the MSTS and Navy is alluded to in quoted material at the Preface head, and on the preceding page, in which Captain Grabowsky describes his experiences commanding Task Group 73.5 in the Tonkin Gulf. Responsibility for the twenty to thirty oilers, supply ships, and ammunition ships comprising it at any given time was rotated between commanders of service squadrons on the U.S. West Coast and Pearl Harbor and one of the staff officers from the staff of the Seventh Fleet. The use of MSTS ships as floating forward depots to support Seventh Fleet's ammunition requirements, required the transfer of remaining stocks aboard the AEs before the ammunition ships departed for Subic to reload. The transfer at sea of large quantities of ammunition to merchant ships was dangerous, but saved the Navy valuable time and money.[6]

Less dangerous than duty aboard a "floating ammunition dump" off Vietnam, but still disconcerting to some, was assignment to an oiler or tanker carrying hundreds of thousands of gallons of volatile fuel. Six relatively small, 310-foot gasoline tankers regularly went in harm's way to deliver fuel desperately needed by combat forces ashore.

Photo Preface-5

USS *Elkhorn* (AOG-7) under way on 1 February 1968, and a ship's plague from 1966. Naval History and Heritage Command photographs #NH 84951 and NH 75739-KN

Elkhorn, *Genesee*, *Kishwaukee*, *Noxubee*, *Patapsco*, and *Tombigbee* were units of Service Squadron Five, based in Hawaii. From March 1965 until late 1971, at least one AOG was constantly deployed in Vietnamese waters, during which their job was to deliver petroleum products to the outposts of I Corps. Bordering the Demilitarized Zone to the north, I Corps saw heavy fighting almost continuously from 1956 to 1975.[7]

Map Preface-2

Abbreviated map of South Vietnam, showing I-Corps, and part of II-Corps of the five total Corps Tactical Zones

Upon arrival off a shore station to deliver fuel, AOGs normally anchored 1,500 to 2,000 yards offshore, and connected hoses to the seaward terminus of a pipeline laid on the seafloor. During typhoon season, churned waters caused breaks in pipelines. Until repairs could be made, tankers had to move in very close to transfer fuel ashore. In an effort to destroy the lifeblood needed by allied patrol craft, vehicles, and aircraft to sustain combat operations, the North Vietnamese regularly took tank farms under fire with shore artillery. In such cases, any AOG present was also an attractive target.

A Marine logistics base located at the mouth of the Cua Viet River was the most dangerous stop for the gasoline tankers along their routes. Collocated with the Naval Support Activity, Da Nang, Cua Viet detachment, the base lay just five miles south of the Demilitarized Zone separating North and South Vietnam. In one of many such attacks, in April 1968, North Vietnamese Army artillery fire hit the base's fuel farm, destroying 40,000 gallons of petroleum. Its loss required immediate delivery of additional fuel by an AOG.

TUGS, SALVAGE, AND RESCUE SHIPS (Thirty-five)

The next largest category of Service Force ships was the fleet tugs, auxiliary tugs, salvage ships, and rescue vessels. Constant salvage work requirements in Vietnam began in the summer of 1965 and continued for several years.

Salvage Force Ships

ARS Salvage Ships - 9
Bolster (ARS-38), *Conserver* (ARS-39), *Current* (ARS-22), *Deliver* (ARS-23), *Grapple* (ARS-7), *Grasp* (ARS-24), *Opportune* (ARS-41), *Reclaimer* (ARS-42), *Safeguard* (ARS-25)

ASR Submarine Rescue Vessels - 4
Chanticleer (ASR-7), *Coucal* (ASR-8), *Florikan* (ASR-9), *Greenlet* (ASR-10)

ATA Auxiliary Ocean Tugs - 4
Mahopac (ATA-196), *Sunnadin* (ATA-197), *Tillamook* (ATA-192), *Wandank* (ATA-204)

ATF Fleet Tugs - 17
Abnaki (ATF-96), *Apache* (ATF-67), *Arikara* (ATF-98), *Chowanoc* (ATF-100), *Cocopa* (ATF-101), *Hitchiti* (ATF-103), *Lipan* (ATF-85), *Mataco* (ATF-86), *Moctobi* (ATF-105), *Molala* (ATF-106), *Munsee* (ATF 107), *Quapaw* (ATF-110), *Shakori* (ATF-162), *Sioux* (ATF-75), *Tawakoni* (ATF-114), *Tawasa* (ATF-92), *Ute* (ATF-76)

ANL Net Laying Ship - 1
Cohoes (ANL-78)

The most unique member of the above group was the net layer *Cohoes* (ANL-78), assigned to Harbor Clearance Unit One at Subic Bay in the Philippines. Established on 1 February 1966, HCU-One was based at Subic to allow salvage operations in both Vietnamese waters and throughout the Pacific.[8]

The 168-foot net laying ship had been commissioned on 23 March 1945 to protect US Navy ships and harbors. Decommissioned on 3 September 1947, she was laid up in the Pacific Reserve Fleet until brought out of "mothballs" in 1968 to serve as a harbor and river clearance craft. Modifications made to prepare her for this new role included increased bow lift capability, allowing her to handle beach gear, and installation of a divers' air system.[9]

Photo Preface-6

Net laying ship USS *Cohoes* (ANL-78) under way on 6 May 1968, and ship's patch. National Archives photograph #USN 1132272

During duty in the combat zone between 3 July 1968 and 3 April 1972, *Cohoes* was assigned to U.S. Naval Support Activity, Da Nang. She and her small ship's complement (4 officers and 45 men) were awarded the Meritorious Unit Commendation multiple times, and also a Navy Unit Commendation and Combat Action Ribbon. Justification for these awards may be found in Appendix A, which includes a reference to the incident associated with the Combat Action Ribbon:

> While under heavy enemy artillery attack, COHOES made an extremely difficult but successful salvage of a patrol craft sunk in the Cua Viet Channel. COHOES' speedy clearance of the channel was of great assistance in the movement of vitally needed cargo. In addition, she successfully salvaged two barges and towed free a lighter which had run aground, all within an eight-day period.

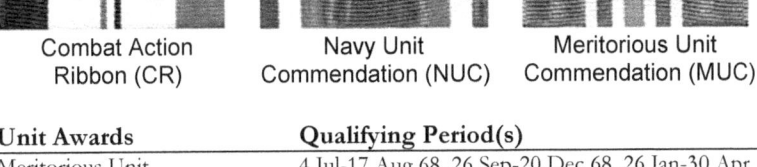

Combat Action Ribbon (CR) Navy Unit Commendation (NUC) Meritorious Unit Commendation (MUC)

Unit Awards	Qualifying Period(s)
Meritorious Unit Commendations:	4 Jul-17 Aug 68, 26 Sep-20 Dec 68, 26 Jan-30 Apr 69, 11 Jun-31 Jul 69, 21 Aug-8 Oct 69, 7 Nov-18 Dec 69, 31 Jan-19 Mar 70, 29 Sep-15 Nov 70, 1 May 71-1 Apr 72
Navy Unit Commendation:	1 Jul 70-30 Jun 71
Combat Action Ribbon:	3 Aug 1968

It must be pointed out that qualifying periods for unit awards did not necessarily mean receipt of that number of awards. Some awards stood alone; in other cases, the awarding authority might sequentially add additionally qualifying periods to the same award. (Appendix B provides a comprehensive list of the Combat Action Ribbons, Navy Unit Commendations, and Meritorious Unit Commendations earned by the 135 Service Force ships.)

BOLSTER WITNESSES THE MOST COMBAT ACTION

Salvage ship *Bolster* shared something in common with *Cohoes*, and gasoline tankers earning combat action ribbons—Cua Viet. As previously noted, the detachment at Cua Viet, from Naval Support Activity, Da Nang, was within range of North Vietnamese artillery and subject to frequent attacks. Eight miles up the Cua Viet River, from Cua Viet at its mouth, was another detachment at Dong Ha, which served as one of the main supply points for forces operating in such places as Con Thien and Quang Tri. The North Vietnamese wanted to deny allied forces the fuel delivered to Cua Viet, and fuel and other combat materiel ferried by craft up the snaking river to Dong Ha.

In February 1968, following the Tet Offensive, begun by the North Vietnamese and Viet Cong prior to dawn on 30 January across South Vietnam, the Dong Ha/Cua Viet area received twenty-six separate rocket/artillery attacks during the month. Twenty-seven craft from Naval Supply Activity, Da Nang, were hit at these and other locations with varying degrees of damage as a result. *Bolster* came under enemy fire at Cua Viet on 7, 16, and 21 February 1968, earning three of her four combat action ribbons. She garnered a fourth on 4-5 May 1970.[10]

Photo Preface-7

Diver alongside the salvage ship *Bolster* (ARS-38) off the coast of Vietnam, July 1969; and *Bolster* under way in June 1974.
National Archives photographs #USN 1139404 and #DN-SC-86-00181

On each of these occasions, *Bolster* was at Cua Viet to conduct salvage operations on damaged craft, or repair POL lines or moorings used by tankers to transfer petroleum products ashore to storage facilities. Fortunately, there were no casualties and only one crewman injured while under fire. On 21 February, *Bolster* received rounds from the beach close aboard to port, during which CS3 (DV) James W. Mott Jr., USN, lacerated his right shin when he fell while seeking cover. Following medical treatment, the Commissaryman Third was returned to all duties except for diving.[11]

Photo Preface-8

Bolster's dive boat towing a repaired four-inch fuel line into position. The line will rest on the sea floor for tanker transfer of fuel ashore. Off the coast of Vietnam, July 1965. National Archives photograph #USN 1139765

SERVICE FORCE, PACIFIC, ORGANIZATION

Photo Preface-9

Rear Adm. Edwin B. Hooper, USN, commander, Service Force Pacific (left); and (at right), Rear Adm. Norvell G. Ward, USN, commander, Service Group Three. USS *Tutuila* (ARG-4) 1967-1968 cruise book

Commander, Service Force, U.S. Pacific Fleet, coordinated the actions of the logistic ships and shore support facilities throughout the Pacific area, including supplying the Navy in Southeast Asia. Principal subordinate commands were Service Group One, based in San Diego for the Eastern Pacific; Service Squadron Five in Hawaii; and Service Group Three, at Sasebo, Japan, for the Western Pacific. Commander, Service Squadron Three had both administrative and operational roles, being "double-hatted" as commander, Task Force 73 (responsible for the Seventh Fleet's logistic Support Force).[12]

The flexible and versatile task force could concentrate a great number of ships in Southeast Asia to provide units of the deployed fleet with ammunition, fuel, supplies, and repairs. In addition to the services discussed thus far, the task force also provided communications, towing, port services, postal and medical support, as well as the universally-desired movies that passed from ship to ship. Continuous support for the fleet enabled ships to operate for long periods at Yankee and Dixie stations, on Market Time patrol (to interdict the flow of arms, ammunition, and other war materiel into South Vietnam), and on the gunline off Vietnam.[13]

ATLANTIC FLEET SHIPS ALSO SENT TO VIETNAM

Demand for Service Force ships in Vietnam also resulted in some Atlantic Fleet units being deployed to the Western Pacific. Transit down the eastern seaboard of the United States and passage through the Caribbean and Panama Canal, preceded a long, and sometimes rough ocean crossing to Southeast Asia.

Photo Preface-10

Pacific crossing in 1972 aboard *Mount Baker*.
USS *Mount Baker* (AE-29) 1972 cruise book

REMAINING SERVICE FORCE SHIPS (Twenty-eight)
Smaller numbers of a variety of other type ships comprised the balance of the Service Force that served in Vietnam. Former sailors would likely expect this category to include repair ships, cargo and general stores ships, and perhaps a hospital ship or two. The activities of the other ships—major communications relay ships, survey ships, seaplane tenders, and particularly the technical research (intelligence-gathering) ships—were less visible, and thus lesser known, to the fleet.

Tenders and Repair Ships (nine)
AD   Destroyer Tenders - 3
Isle Royale (AD-29), *Piedmont* (AD-17), *Samuel Gompers* (AD-37)
AR   Repair Ships - 6
Ajax (AR-6), *Delta* (AR-9), *Hector* (AR-7), *Jason* (AR-8), *Klondike* (AR-22), *Markab* (AR-23)

Cargo Ships (four)
AKL   Light Cargo Ships – 2
Brule (AKL-28), *Mark* (AKL-12)
AKS   General Stores-Issue Ships - 2
Castor (AKS-1), *Pollux* (AKS-4)

Miscellaneous Ships (ten)
AGMR   Major Communications Relay Ships – 2
Annapolis (AGMR-1), *Arlington* (AGMR-2)
AGS   Survey Ships – 6
Maury (AGS-16), *Rehoboth* (AGS-50), *Serrano* (AGS-24), *Sheldrake* (AGS-19), *Tanner* (AGS-15), *Towhee* (AGS-28)
AGTR   Technical Research Ships – 2
Jamestown (AGTR-3), *Oxford* (AGTR-1)

Aviation Support Ships (three)
AV   Seaplane Tenders -3
Currituck (AV-7), *Pine Island* (AV-12), *Salisbury Sound* (AV-13)

Other Service Force Ships (two)
AH   Hospital Ships - 2
Repose (AH-16), *Sanctuary* (AH-17)

A majority of the service ships that served in Vietnam were old. Used up by service in World War II and the Korean War, they'd been laid up in backwater Reserve Fleets, should the nation again require their services. The Vietnam War spurred such a need. Brought out of "mothballs" and pressed into high tempo operations, some "clapped out" vessels proved challenging to keep running. One such, apparently, was the technical research ship USS *Jamestown* (AGTR-3), based on a poem describing duty aboard her. The first line makes reference to the aircraft carrier USS *Kitty Hawk* (CVA-63).

THE JIMMY-T

She hasn't the power of the Hawk named Kitty,
The oilers outrun her and she's not very pretty,
She's never been called a ruler of the sea,
But she's all we've got, she's the Jimmy-T

Her engines drink oil with unquenchable thirst,
She's got sick generators which have to be nursed,
Her paint is as thick as the bark on a tree.
But she's all we've got, she's the Jimmy-T

We curse and we cuss and we rant and we rave,
About chow, or pay, or how we all slave,
We'll work it all off on the beach with a spree,
Then stagger back home to the ole Jimmy-T.

—Author unknown; courtesy of Bob Harper,
former USS *Jamestown* crewmember (1968-1969)

SERVICE FORCE SHIPS IN HARM'S WAY

Despite such challenges for some of its members, the Service Force acquitted itself well providing support for the fleet, during which a dozen ships earned Combat Action Ribbons.

Name	Survey Ships Combat Action Ribbon Award Period(s)
Maury (AGS-16)	11 December 1965
Serrano (AGS-24)	3-4 March 1967

	Stores Ship
Mars (AFS-1)	17 September 1966

	Light Cargo Ships
Brule (AKL-28)	28 January 1967, 24 August 1968
Mark (AKL-12)	15 February 1967, 19 November 1969

	Salvage Ships
Bolster (ARS-38)	7 February 1968, 16 February 1968
	21 February 1968, 4-5 May 1970
Deliver (ARS-23)	27-28 February 1968

	Gasoline Tankers
Genesee (AOG-8)	22 April 1968
Kishwaukee (AOG-9)	21 February 1968
Noxubee (AOG-56)	28 October 1968, 9 September 1969
Patapsco (AOG-1)	16 February 1968, 27-28 February 1968

	Net Laying Ship
Cohoes (ANL-78)	3 August 1968

RAN SUPPORT FOR THE VIETNAM WAR

As detailed in *On the Gunline*, and *Gators Offshore and Upriver*, the Royal Australian Navy deployed forces to Vietnam from 1965 to 1972, in support of the Allied war effort. Primary contributions included destroyers serving on the gunline off the coast, helicopters conducting combat missions in-country, clearance divers inspecting ships at Vung Tau for enemy-emplaced explosives, and a logistic support force of transport, supply, and escort ships.

Summary information about these forces follow, along with identification of merchant vessels which carried vital materiel to Vietnam. For the ships, and other RAN units awarded Battle Honours, the associated year(s) identify the span of their service.[14]

Gunline Destroyers

Ship	Battle Honours	Ship	Battle Honours
HMAS *Brisbane*	1969-71	HMAS *Perth*	1967-71
HMAS *Hobart*	1967-70	HMAS *Vendetta**	1969-70

*Vendetta also served as an escort during the war

Logistic Support

HMAS *Boonaroo*	1967	HMAS *Sydney*	1965-72
HMAS *Jeparit*	1969-72		

Helicopter Flight Vietnam

No. 723 Squadron	1967-71	No. 725 Squadron	1967

Clearance Diving Teams

CDT 1	1966-71	CDT 3	1967-71
CDT 2	1967-71		

Escorts (No Battle Honours)

HMAS *Anzac*	HMAS *Melbourne*	HMAS *Swan*	HMAS *Yarra*
HMAS *Derwent*	HMAS *Parramatta*	HMAS *Torrens*	
HMAS *Duchess*	HMAS *Stuart*	HMAS *Vampire*	

Merchant Vessels (No Battle Honours)

Brudenell White AV 1354	*Harry Chauvel* AV 1353	*Vernon Sturdee* AV 1355
Clive Steel AV 1356	*John Monash* AS 3051	

Chapters 4-5 of this book describe the activities of the logistic support ships HMAS *Sydney*, HMAS *Boonaroo*, and HMAS *Jeparit*, and their contributions to the war effort. Chapter 15 describes two salvage operations carried out by Clearance Diving Team 3. During the RAN's involvement in the war, eight CDT 3 contingents deployed to Vietnam in succession, each comprised of one officer and five enlisted men. The two operations were under Lt. Michael T. E. Shotter, RAN (First Contingent) and Lt. Edward ("Jake") W. Linton, BEM RAN (Eighth Contingent).

IDENTIFICATION OF THE 135 SERVICE FORCE SHIPS

Before readers delve into the body of the book, an explanation of the number of ships it's devoted to, might be in order. Following its authorization by President Lyndon B. Johnson, the Vietnam Service medal was awarded from 4 July 1965 to 28 March 1973, and for the evacuation of Saigon from 29-30 April 1975. Prior to the establishment of this medal, ships and their crews were eligible for receipt of the Armed Forces Expeditionary Medal from 1 July 1958 through 3 July 1965. The 135 ships to which *Support for the Fleet* is devoted, all served one or more tours on the line between 4 July 1965 and 28 March 1973.

A handful of ships, which might rightly be considered Service Force vessels, are not included in the book because they were primarily a part of the "in-country" forces. These include the internal combustion engine repair ship *Tutuila* (ARG-4), and non-self-propelled barges and the tugs that moved them, assigned to the Mobile Riverine Force.

With this introduction to the book completed, it's time to stand out to sea with the men and women of the U.S. Navy's Service Force.

1

Ambush on the Co Chien River

> *Ship and crew ready to haul cargo or fight and not necessarily in that order.*
>
> —Statement in a damage report submitted by USS *Brule* AKL-28 (the former Army freight-supply ship *FS-370*), after sustaining rocket hits in August 1968 during one of her normal supply runs in inland waters of South Vietnam. Despite suffering extensive damage to her superstructure and cabling, she suppressed the enemy fire and proceeded on schedule.[1]

Photo 1-1

Painting by Richard DeRosset of the light cargo ship USS *Brule* (AKL-28) under attack by rocket and automatic weapons fire coming from both banks of the Co Chien River. She was ambushed by the Viet Cong on 24 August 1968 while proceeding downriver.

On 24 August 1968, as the dawn of another hot and humid day broke in Vietnam, USS *Brule* lay at anchor in the Co Chien River off Vinh Long, with various units of the River Patrol Force. Her sea detail was set at 0715, and the light cargo ship was soon under way, maneuvering on various courses and speeds to conform to the channels in the river. On the bridge, her commanding officer, Lt. William Hewitt, USN, "had the conn" (personal control of the movement of the ship via rudder and engine commands). Hewitt was a "Mustang," a former chief signalman who had earned a commission as a Deck, Limited Duty Officer. Propelled by two 500hp GM 6-278A diesel engines, *Brule* had a modest top speed of 13 knots. Her tasking that day was to resupply the tank landing ship USS *Garrett County* (LST-786), located down the river. The water's surface was smooth that morning.[1]

Photo 1-2

USS *Brule* (AKL-28) at anchor in South Vietnamese waters, 27 September 1966. National Archives photograph #USN 1118482

Garrett County (LST-786) was one of four World War II-vintage tank landing ships serving as mobile support bases for river patrol boats (PBRs) and "Seawolf" helicopter gunships in support of friendly forces ashore and along the rivers of the Mekong Delta. She, *Harnett County* (LST-821), *Hunterdon County* (LST-838), and *Jennings County* (LST-846) were a part of Task Force 116 (River Patrol Force), engaged in Operation GAME WARDEN. This code word referred to using 31-foot PBRs, assisted by armed helicopters, to limit the enemy's use of the larger waterways in South Vietnam. This was necessary to interdict supplies for the Viet Cong flowing into the Mekong Delta from Cambodia.

Helicopter Attack Squadron (Light) Three, Detachment 4, and River Patrol Section 523 were then operating from the *Garrett County*.[2]

Photo 1-3

USS *Garrett County* (LST-786) on the Co Chien River, Mekong Delta, in June 1968, while providing support for five PBRs and two UH-1Bs of Operation GAME WARDEN. National Archives photograph #K-51442

PEDIGREE OF *BRULE* AND HER SISTER SHIPS

Brule and sister ship *Mark* were 176-foot, former unnamed Army freight supply ships (the *FS-370* and *FS-214*). In World War II, the U.S. Army—requiring vast numbers of small ships to support its ground forces in the Southwest Pacific, and in other theaters of war—had contracted for the construction of hundreds of freight and freight-supply ships. Following the war, between 1947 and 1966, the Navy acquired forty-five of the rugged steel, war-tested freighters. The vessels were originally commissioned Miscellaneous Auxiliary Ships, and a few years later were reclassified, on 1 July 1950, as Light Cargo Ships. Nine of these ships received unit awards for combat duty in the Korean or Vietnam War, or for service as Environmental Research ("intelligence gathering") ships.[3]

The most well-known is the USS *Pueblo* (AGER-2), captured by North Korean naval forces on 23 January 1968. She remains today a commissioned United States Navy Ship, despite being held by the Communist country. The former freight-supply ship most viewed by the public was the light cargo ship USS *Hewell* (AKL-14). She garnered seven battle stars during the Korean War, but is best known as the fictitious Navy cargo ship *Reluctant* depicted in the 1955 American comedy-drama film *Mr. Roberts*. Filmed in Hawaii, the movie stared Henry Fonda, James Cagney, William Powell, and Jack Lemon. The film was based on a novel and screenplay by Thomas Heegan, who served as communications officer aboard the ammunitions ship USS *Virgo* (AE-30) during World War II (discussed in Chapter 18).[4]

Other freight-supply ships also had interesting post-World War II service. Following her naval stint as USS *Deal* (AKL-2), the former unnamed *FS-263*, operated in 1966 off the coast of England as a pirate radio station transmitting "Swinging Radio England" (SRE), initially as the motor vessel MV *Olga Patricia* and, after a name change, as the *Laissez Faire*. As depicted in the 2009 motion picture *Pirate Radio*, from aboard the ex-*Deal* and other ships located in international waters off England's east coast, rebellious disc jockeys were broadcasting rock' n' roll music which, although spreading like wildfire in the United States, was all but banished from the British airwaves. The BBC owned all but one commercial TV network, and the broadcasting corporation favored a bland fare of news and information, light entertainments and children's programs. More recently, the former freight-supply and "radio pirate" ship was employed as the fishing vessel *Earl J. Conrad Jr.*, out of Reedville, Virginia.[5]

USS *Mark* (AKL-12) and *Brule* (AKL-28) garnered four combat action ribbons, five Navy Unit Commendations, and thirty-two other unit awards between them for Vietnam War service. A summary of the awards received by them and seven other former freight-supply ships for Korean War service and beyond, follows.

	Korean War				
	ex *FS-263* USS *Deal*	ex *FS-275* USS *Estero*	ex *FS-361* USS *Ryer*	ex *FS-385* USS *Sharps*	ex *FS-263* USS *Hewell*
Battle Stars	7	7	6	3	7

	Vietnam War	
	ex *FS-370* USS *Brule*	ex *FS-214* USS *Mark*
Combat Action Ribbons	2	2
Navy Unit Commendation	2	3
Vietnam Service Medals	11	14
Republic of Vietnam Meritorious Unit Citations, Gallantry	3	4

	Intelligence Gathering Missions	
	ex *FS-345* USS *Banner*	ex *FS-344* USS *Pueblo*
Combat Action Ribbon		1
Armed Forces Expeditionary Medal	1	1

Following their acquisition by the Navy, the *Brule* and *Mark* were placed in commission as AKLs, and later from 1956 to 1965 placed "out of commission in service" and used to shuttle supplies and passengers between Subic Bay and U.S. Naval Station, Sangley Point, located on a peninsula jutting into Manila Bay, about eight miles southwest of the city of Manila. *Brule* and *Mark* were re-commissioned on 1 September, and 1 October 1965, respectively, and sent to Vietnam. Attached to Service Group Three, they worked directly for U.S. Naval Support Activity, Saigon, distributing vital cargo and supplies up and down shallow waterways in the Mekong Delta region of Vietnam.[6]

Brule averaged three trips a month from Nha Be/Saigon to ports such as Vung Tau, Cat Lo, My Tho, Dong Tam, Chau Doc, Binh Thuy, An Thoi and other operational bases on the rivers and coast of Vietnam. She transported general cargo, ammunition and POL products: on an average run, up to 225 tons of cargo, 38,000 gallons of water and 20,000 gallons of fuel oil. Of modest size and appearance, she and her small crew of 43 enlisted and 5 officers provided vital support to the "Brown Water Navy" (the name ascribed to the delta force) while fulfilling her valiant motto: "SERVICE – OUR MISSION FOR FREEDOM"[7]

RIVER NARROWS POSE DANGER

Brule, and other ships and craft that plied the rivers, knew the popular ambush sites used by the enemy. These were typically where a waterway narrowed, such as between an island and a river bank, or other locations when it was necessary for a vessel to pass sufficiently close to a bank to come under attack. Making passage along the center of a wide tributary was preferred, as it avoided these situations.

At 0747 on 24 August, only nine minutes after getting under way from Vinh Long, *Brule*'s commanding officer secured the sea detail and set General Quarters. This action proved prudent, because at 0814 the light cargo ship began taking rocket and automatic weapons fire from the starboard bank of the Co Chien River, as she was abreast a brick factory. *Brule* was steering a course down the centerline of the river, or nearly so. One minute later, rocket and small arms fire erupted from the port bank. Her gun crews were on "top form" that morning—the starboard battery quickly opened fire, followed by the port battery, and suppressed the hostile fire at 0816.[8]

Photo 1-4

USS *Brule* (AKL-28) on the Bassac River, South Vietnam, in September 1968. Courtesy of Albert Moore

All mounts then ceased fire and gun crews reloaded. The *Brule*'s armament consisted of two .50-caliber machine guns on her fo'c'sle, and two larger 20mm guns above the wheelhouse—one of each type on the port and starboard sides of the ship. Warrant Officer (WO1) Thomas L. Young, USN, at his gun control station above the wheelhouse, had directed the actions of the two batteries. Observing rocket and automatic-weapons fire coming from the vicinity of big, domed ovens used for baking bricks, he had ordered the starboard battery into action. (The weapons themselves, and enemy wielding them, were shielded from his view.) Knowing that when an attack originated from one side of a ship, one from the other often followed, he shifted his field of vision to *Brule*'s portside, noted weapons fire from the left bank, and ordered the port battery into action as well.[9]

Brule suffered material damage, identified below, but no personnel casualties, owing at least in part to her rapid counterfire:

- Rocket hit, starboard side at frame 77, one foot above the waterline, destroying freon lines in No. 2 cargo hold and severing power cables forward
- Rocket hit, starboard side forward of the pilot house, destroying an educator (pump) hose and freon bottles, and shattering two pilot house portholes

- Rocket hit, port bridge wing, damaging lighting in the forward berthing compartment
- Automatic weapons fire damaged the 1-MC announcing system, CO2 bottles, and lube oil drums, and created numerous small holes throughout the ship[10]

Photo 1-5

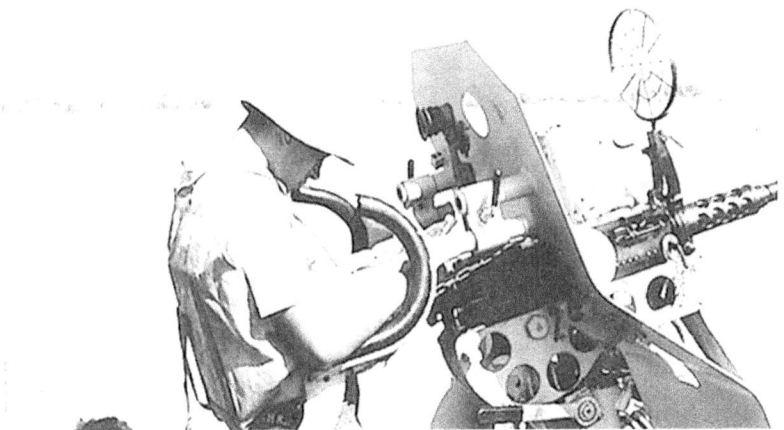

One of USS *Brule*'s two Browning .50-caliber heavy machine guns.
Courtesy of Craig Reynolds, and the Mobile Riverine Force Association

Four of the seven rockets fired at *Brule* missed the ship. Although busy directing the gun action, Young did witness the second to last, and final misses, both fired from the starboard bank. The first of these two rockets disappeared at a 45-degree angle into the sky; the last one hit forward of the ship and skipped across the water.[11]

Lieutenant Hewitt ordered General Quarters secured at 0958 and set the sea detail. Fifteen minutes later, his ship moored port side to the *Garrett County*. After resupplying her, *Brule* cleared the tank landing ship's side, and began transit up the Co Chien and My Tho Rivers to Sa Dec. Arriving at her destination late afternoon, she anchored in the My Tho off the river patrol boat base, ready to continue her duties as one of the units of the Service Force in Vietnam.[12]

Photo 1-6

Crewmembers of the USS *Mark* (AKL-12), sistership of USS *Brule*, offloading supplies for Navy units operating in the Mekong Delta.
Department of Defense photograph K-52148

2

Prelude to the Vietnam War

You can kill ten of my men for every one I kill of yours, but even at those odds, you will lose and I will win.

—Remark by Ho Chi Minh to the French in the late 1940s.

Photo 2-1

In Hanoi, a large parade poster shows Ho Chi Minh, ruler of North Vietnam, shaking hands with party chairman Mao Tse-Tung of Communist China, symbolizing Peking backing of the North Vietnamese Campaign against South Vietnam.
U.S. Information Agency photograph #65-1277

FRENCH INVOLVEMENT IN INDOCHINA

Prior to American involvement in Southeast Asia, the French had long occupied Vietnam and adjacent Cambodia and Laos. It began in

September 1858, with a Franco-Spanish expedition to the city of Tourane (Da Nang). After the subsequent French conquest of Saigon and three southern provinces, the Vietnamese government was forced to cede the southern portion of the country to France in 1862. The resulting French colony was named Cochinchina and, in 1887, France added the balance of Vietnam, as well as Cambodia, to its empire to create the Indochina Union. "French Indochina," as it was commonly known, was later expanded by the accession of Laos in 1893, followed by that of Kouang-Tcheou-Wan, a small enclave on the south coast of China ceded to France as a leased territory, in 1900.[1]

Photo 2-2

Boat landing on the Mekong River, French Indochina, circa 1931.
Naval History and Heritage Command photograph #NH 80423

Japan occupied French Indochina in September 1940, but left the French colonial government intact before taking over administration of the area as a protectorate, near the end of World War II. After recognizing in 1945 that defeat was inevitable, Japan allowed the countries that made up the Indochina Union to proclaim their independence from France. This freedom was short-lived in the case of Laos and Cambodia, which readily accepted the return of the French overlords later that year. Vietnam proved much different, as Ho Chi Minh, the leader of the Indochina Communist Party, was unyielding in his desire to rid the country of foreign dominance. Shortly after the surrender of Japan in August 1945, Ho announced the establishment of a provisional government in Hanoi with Bao Dai as supreme counselor and, a few days later, declared Vietnam's independence.[2]

Ho Chi Minh was born in central Vietnam in 1890 and left the country in 1911 as an adult. During a lengthy absence, he joined the French Communist Party in 1920, and then travelled from Paris to Moscow to become, four years later, a full-fledged Communist agent, and eventually to Hong Kong to form the Indochina Communist Party in 1940. The following year, Ho returned covertly to Vietnam and, from inside China on the Vietnamese border, established the Vietnam Revolutionary League ("Vinh Minh"). Under Gen. Vo Nguyen Giap, the Vinh Minh began a guerrilla campaign against the Japanese that continued until the end of World War II. Following Japan's surrender, British forces landed in Saigon on 13 September 1945, and thereafter returned authority to the French. That November, Ho Chi Minh dissolved the Indochina Communist Party and replaced it with the Association for Marxist Studies in an effort to broaden his base of support.[3]

A few months earlier, the heads of the Soviet Union, the United Kingdom, and the United States had met at Potsdam, Germany, to determine how to administer a defeated Germany following its surrender on 8 May 1945. Goals of the conference included establishing post-war order, resolving peace treaty issues, and countering the effects of war. Following discussion about the Japanese occupation of French Indochina, on 26 July 1945, Harry S. Truman, Winston Churchill, and Chiang Kai-shek issued the Potsdam Declaration calling for the surrender of Japan. In a related decision, the allied leaders determined that the British would disarm the Japanese forces in southern Vietnam, and the Chinese nationalists would perform this function north of the sixteenth parallel.[4]

Following Japan's surrender on 2 September 1945 and the allied disarmament of its troops, China agreed in February 1946 to withdraw its forces from North Vietnam, and the French and Viet Minh reached an accord in March that recognized Vietnam as a "free state" within the French Union. However, any resulting goodwill between the two parties disappeared three months later when Adm. Thierry d'Argenlieu, the French high commissioner for Cochinchina, violated the agreement by proclaiming a separate government for Cochinchina. Hostilities between the two factions resulted and, after negotiations broke down, French warships bombarded Haiphong Harbor and French troops occupied Hanoi, forcing Ho Chi Minh to withdraw his forces from the city and create a rural base from which to operate.[5]

Photo 2-3

French Rear Admiral Thierry D'Angenlieu (in white) walking with (at his left, in blue) French General Richard Brunot, location and date unknown.
Naval History and Heritage Command photograph #NH 121080

The French Indochina War, fought from 19 December 1946 to 1 August 1954, between the French Union's French Far East Expeditionary Corps (led by France and supported by Bao Dai's Vietnamese National Army) and the Viet Minh (led by Ho Chi Minh and Gen. Vo Nguyen Giap) had begun. Most fighting took place in Tonkin, the northern part of Vietnam, although the conflict engulfed parts of the entire country and extended into the neighboring French Indochina protectorates of Laos and Cambodia.[6]

After seven long years, with Laos and Cambodia demanding their independence and a majority of the French National Assembly expressing a desire for a negotiated end to the war, the French government finally granted the protectorates their freedom. In 1941, the French enthroned Prince Norodom Sihanouk, at the age of eighteen, as the ruler of Cambodia, believing him easily controlled. This assumption would prove false. In 1953, after having increasingly demanded full independence of the kingdom from the French colonists and their departure from Cambodia, Sihanouk embarked on a world tour to publicize his campaign. Fearing retribution for this action, he afterwards went into voluntary exile in Battambang Province, joined by 30,000 Cambodian troops and police. The French government granted Cambodia its independence on 9 November 1953. It had made the same concession to Laos the preceding month.[7]

FRENCH DEFEAT AT DIEN BIEN PHU

> [Dien Bien Phu] was the first time that a non-European colonial independence movement had evolved through all the stages from guerrilla bands to a conventionally organized and equipped army able to defeat a modern Western occupier in pitched battle.
>
> —Martin Windrow, military historian quoted in the *Boston Globe*, 4 January 2005.

The final blow to France's tenuous hold on what remained of the Indochina Union was delivered by Viet Minh communist revolutionaries in a stunning defeat of the French Far East Expeditionary Corps at the Battle of Dien Bien Phu. The battle took place from 13 March to 7 May 1954 in northwestern Vietnam near the Chinese and Laotian borders. The Vo Nguyen Giap-led Viet Minh occupied the highlands surrounding the battlefield, besieged the French with heavy artillery fire and, after a lengthy siege, overran their garrison and killed or captured most of the forces. Of the French soldiers captured, few survived the ensuing grim death march to Viet Minh prison camps located 300 miles to the east.[8]

The Battle of Dien Bien Phu was the defining conflict of the French Indochina War, and the crushing French defeat influenced negotiations over the future of Indochina. Agreements reached at the Geneva Convention, in July 1954, called for the cessation of hostilities in Vietnam, Cambodia, and Laos, and temporarily divided Vietnam at the seventeenth parallel, pending elections in 1956 to choose a national government that would administer a reunified country. Until then, the Viet Minh would remain in charge of North Vietnam, while the State of Vietnam (which succeeded the Provisional Central Government of Vietnam that existed from 1948-1949) controlled the South, and French forces gradually withdrew from Vietnam as the situation stabilized. Bao Dai, the former emperor of Vietnam (8 January 1926-25 August 1945), was the State of Vietnam's chief of state, and Ngo Dinh Diem its prime minister. Neither the United States nor the State of Vietnam signed the Geneva Accords. The United States indicated that it would not disturb the agreements, but would view renewed aggression with concern. Ngo Dinh Diem rejected the idea of a nationwide election, believing that a free open-ballot vote was impossible in the communist North.[9]

Photo 2-4

Refugees board the tank landing ship USS *LST-516* for their journey from Haiphong to Saigon, October 1954, as part of Operation PASSAGE TO FREEDOM. National Archives photograph #80-G-652364

The accords also stipulated a period of grace ending on 18 May 1955, in which people could move freely between the two Vietnams before the border at the seventeenth parallel was sealed. During this period, an estimated 310,000 Vietnamese civilians, soldiers, and non-Vietnamese members of the French Army fled to the south by sea as part of the U.S. Navy Operation PASSAGE TO FREEDOM. In total, the cargo vessels and tank carriers that participated in the evacuation as part of Task Force Ninety would make 109 southbound voyages carrying their charges to freedom. The first vessel to embark refugees was the *Menard* (APA-201) in the fall of 1954. Other task force units included the *Bayfield* (APA-33) and *Montague* (AKA-98), and the USNS *General A. W. Brewster* (T-AP-155). The French Navy and Air Force transported another 500,000 refugees.[10]

U.S. NAVY'S GROWING ROLE IN VIETNAM

Photo 2-5

Rear Adm. Aaron P. Storres, USN; an unidentified civilian; Lt. Gen. John W. O'Daniel, USA, chief of MAAG (Military Assistance Advisory Group) Vietnam; Rear Adm. Lorenzo S. Sabin Jr., USN; and U.S. Ambassador to Vietnam Donald R. Heath await arrival in Saigon of the 100,000th refugee, September 1954.
National Archives photograph #80-G-647030

The U.S. Navy's initial advisory effort in Vietnam began in October 1950 with the establishment of a Navy Section, within the U.S. Military Assistance and Advisory Group (MAAG), staffed by eight officers and enlisted men. The primary mission of the Navy Section was to provide aid that would help strengthen the coastal and river security of Vietnam. This effort gradually expanded as the section supervised the transfer of craft and equipment to the Vietnamese Navy; furnished military advisors to training centers, the Saigon shipyard, and the coastal and river patrol units; and provided assistance to the Vietnamese Marine Corps.[11]

The U.S. Military Assistance Command Vietnam (MACV) was established on 8 February 1962, following a significant increase in the number of American military personnel in Vietnam. Located at the Tan Son Nhut Air Base outside Saigon, the unified command was responsible for American military activities in Vietnam, while the MAAG continued to administer the military aid program. The functions of the MAAG were absorbed by the MACV in May 1964, and the former Navy Section of the MAAG became the Naval Advisory Group under the U.S. Military Assistance Command. On 10 May 1965, Rear Adm. Norvell G. Ward, USN, assumed the duties of chief, Naval Advisory Group, and on 1 April 1966, the new and concurrent title and responsibilities of commander, Naval Forces Vietnam.[12]

Photo 2-6

Adm. Thomas H. Moorer (chief of Naval Operations designate), at left, is greeted by Gen. William C. Westmoreland, USA, commander, U.S. Military Assistance Command Vietnam (MACV), and Rear Adm. Kenneth L. Veth, commander, U.S. Naval Forces Vietnam (NavForV), on his arrival in Saigon on 27 June 1967.
Naval History and Heritage Command photograph #NH 104909

3

A Number of Firsts in 1965

We can never again stand aside, prideful in isolation. Terrific dangers and troubles that we once called "foreign" now constantly live among us. If American lives must end, and American treasure be spilled, in countries we barely know, that is the price that change has demanded of conviction and of our enduring covenant.

—Excerpt from President Lyndon B. Johnson's Inaugural address, from the east front of the Capitol Building on 20 January 1965.[1]

On 27 January 1965, National Security Advisor McGeorge Bundy and Defense Secretary Robert McNamara, send a memo to President Lyndon B. Johnson advising him that America's limited military involvement in Vietnam was not succeeding, and that the U.S. had reached a 'fork in the road' in Vietnam, and must soon either escalate or withdraw. Early the following month, Johnson approved Operation FLAMING DART, the bombing of North Vietnamese barracks and staging areas near Dong Hoi and Cahn Hoa by aircraft from the attack carriers *Coral Sea* (CVA-43), *Hancock* (CVA-19), and *Ranger* (CVA-61).[2]

This action followed an attack by the Viet Cong 409th Battalion on Camp Holloway (a U.S. Army helicopter base near Pleiku, in the central highlands) on the night of 6-7 February, killing eight Americans, wounding 126 and destroying ten aircraft. Opinion polls of American sentiment after the bombing indicated a 70 percent approval rating for the President and an 80 percent approval of U.S. military involvement in Vietnam. Johnson then agreed to a long-standing recommendation from his advisors for a sustained bombing campaign against North Vietnam. On 22 February, Gen. William C. Westmoreland, USA, commander, U.S. Military Assistance Command Vietnam (MACV), requested two battalions of U.S. Marines to protect the American air base at Da Nang from 6,000 Viet Cong massed in the vicinity. The President approved this request.[3]

Photo 3-1

Camp Holloway, South Vietnam.
DOD photograph

On 2 March 1965, Operation ROLLING THUNDER commenced as American fighter-bombers attacked targets in North Vietnam. Scheduled to last eight weeks, Rolling Thunder would, instead, continue for three-and-a-half years into October 1968. The campaign sought to strike targets of sufficient value to pressure the North Vietnamese into concessions, but which would not result in too many civilian deaths, the destruction of the North Vietnamese regime, or Soviet or Chinese intervention. Therefore, interdiction targets were mainly chosen. These included bridges and railway lines, POL (petroleum, oil, lubricants) facilities, power plants, and weapons and ammunitions storage depots. Absent from the list were many air defense targets deemed too close to heavily populated areas or, in the case of airfields, too provocative.[4]

Photo 3-2

An A-4 Skyhawk prepares to launch from USS *Coral Sea* (CVA-43) on 24 March 1965.
National Archives photograph #USN 1111691

U.S. MARINES LAND AT DA NANG

> *It seems clear that our national policy towards SVN [South Vietnam] is shifting from one in which we attempted to maintain an 'advisory' image in SVN to one of active and overt U.S. participation.*
>
> —Observation by Adm. Thomas H. Moorer, commander, Pacific Fleet, to chief of Naval Operations, Adm. David L. McDonald. Moorer also conveyed to McDonald that his fleet was "on the scene with the capacity and ... are ready to go."[5]

Photo 3-3

U.S. Marines wade ashore from landing craft at Da Nang, South Vietnam. National Archives photograph #USN 1142247

INCEPTION OF OPERATION MARKET TIME

The following month, Task Force 115 (Coastal Surveillance Force) was established on 30 April 1965 to stem the infiltration of weapons, ammunition, and other war materiel into South Vietnam by sea. In support of Operation MARKET TIME, the U.S. Navy deployed U.S. Navy and U.S. Coast Guard destroyer escorts, ocean minesweepers, cutters, coastal craft, and patrol planes along the country's 1,200-mile coastline.[6]

Ten weeks earlier, a North Vietnamese steel-hulled trawler had been located on 16 February at Vung Ro Bay in central Vietnam. Filled with arms and ammunition, its discovery revealed that seaborne infiltration of supplies from North Vietnam into South Vietnam was taking place and that a counter-operation was required. The U.S. Seventh Fleet initiated coastal surveillance operations on 15 March, but soon transferred this responsibility to the Naval Advisory Group, subordinate to MACV.[7]

Photo 3-4

Painting *Operation Market Time* by Gene Klebe, 1965, depicting the inspection of an indigenous craft in South Vietnamese coastal waters. Naval History and Heritage Command accession #88-162-K

Late in the year, as war materiel flowed into the Mekong Delta from Cambodia, Task Force 116 (River Patrol Force) was established on 18 December 1965. To interdict these supplies, 31-foot river patrol boats (PBRs), assisted by armed helicopters, were charged with plying South Vietnam waterways to limit the enemy's use of larger rivers. The code name of this operation was GAME WARDEN.[8]

U.S. ARMY COMBAT TROOPS ARRIVE IN-COUNTRY

On 1 April, President Johnson authorized sending two more Marine battalions and up to 20,000 logistical personnel to Vietnam. Less than three weeks later, Johnson's top aides met on 20 April in Honolulu, and recommended to the president sending another 40,000 combat soldiers to Vietnam.[9]

On 3 May 1965, the first U.S. Army combat troops, 3,500 men of the Okinawa-based 173rd Airborne Brigade, arrived in Vietnam. Nicknamed "Sky Soldiers," they served there from 1965 to 1971. Air Force transports moved the bulk of the troops by air to the base at Bien Hoa, northeast of Saigon. The unit's artillery, anti-tank guns, engineering equipment, and headquarters staff travelled aboard the

Military Sea Transportation Service ship USNS *General William A. Mann* and three MSTS tank landing ships; arriving at Vung Tau, following a five-day voyage from Okinawa.[10]

Photo 3-5

Paratroopers of the 173rd Airborne Brigade arriving at Bien Hoa, South Vietnam. Courtesy of Lew Smith (USASCVIO photograph by Capt. Don Adams)

Photo 3-6

Sky Soldiers of the 2nd Battalion, 503rd Infantry Regiment, 173rd Airborne Brigade en route from Okinawa to Vietnam, aboard the USNS *General William A. Mann* (T-AP-112) in May 1965.
Courtesy of Jim Dresser

Other firsts followed in succession. On 16 May, the destroyer USS *Henry W. Tucker* (DDR-875) fired the first naval gunfire-support mission since the Korean War, against Viet Cong positions near Thang Hai. The following month, USS *Oriskany* (CVA-34) arrived off South Vietnam on 20 May. She would be the first carrier to operate from Dixie Station. On 2 June, USS *Canberra* (CA-70) became the first cruiser to fire 8-inch guns in combat since 1953.[10]

Photo 3-7

Destroyer USS *Henry W. Tucker* (DDR-875) under way, 18 September 1961. Naval History and Heritage Command photograph #NH 106998

PRESIDENT ORDERS INCREASING NUMBERS OF TROOPS AND OTHER FORCES SENT TO VIETNAM

During a noon press conference on 28 July 1965, President Johnson announced that he was sending forty-four combat battalions to Vietnam, thereby increasing the U.S. military presence to 125,000 men, and doubling the number of monthly draft calls to 35,000. In explaining this decision, he stated:

> I have asked the commanding general, General Westmoreland, what more he needs to meet this mounting aggression. He has told me. And we will meet his needs. We cannot be defeated by force of arms. We will stand in Vietnam.
>
> ...I do not find it easy to send the flower of our youth, our finest young men, into battle. I have spoken to you today of the divisions and the forces and the battalions and the units, but I know them all, every one. I have seen them in a thousand streets, of a hundred towns, in every state in this union—working and laughing and building, and filled with hope and life. I think I know, too, how their mothers weep and how their families sorrow.[11]

By year's end in 1965, U.S. troop levels in Vietnam had reached 184,300. Additionally, there were more than 8,000 Navy and Coast Guardsmen in-country; and another 24,000 Navy personnel aboard ships operating off the coast.[12]

4

The "Vung Tau Ferry"

Oh, they loved it [shipboard life]. Well there was a movie every night for them; they got a beer issue, the PTIs [physical training instructors] got them up on the flight deck every morning and gave them their exercises, and then the gunnery people got them down the aft end for rifle practice or gunnery practice, yeah, they thought it was Christmas.

[When homeward troops came aboard,] they were very gaunt, very quiet. They'd do a lot of staring; didn't tend to join in, sort of kept to themselves a lot. That was my first experience of guys that had been in battle and how it really affected them – you can just tell it in their face, you know. Everything was sad about them, you know?

—Bill Kane, a former crewman aboard HMAS *Sydney*, describing transporting fresh Australian troops to Vietnam and, later, returning battle-weary and injured soldiers home.[1]

Photo 4-1

Light fleet aircraft carrier HMAS *Sydney* in Korean waters, circa 1950-1953. Australian War Memorial photograph 044798

The Johnson administration, not wanting to be seen as going it alone, persuaded allies to join in the Vietnam War. South Korea was the principal U.S. and South Vietnam partner, contributing over 300,000 troops and suffering some 5,000 deaths. Nearly 60,000 Australians served (521 of whom died) and over 3,000 New Zealanders (37 killed). Three other countries—the Philippines, Taiwan, and Spain—also aided the U.S. war effort.[2]

AUSTRALIA'S ENTRY INTO THE VIETNAM WAR

On 29 April 1965, a decision was made by the Australian Government to deploy an infantry battalion to South Vietnam. *Sydney* received orders to make preparations to transport the 1st Battalion, Royal Australian Regiment (1RAR) to Vung Tau. With 347 officers and men of 1RAR embarked, along with vehicles and stores, she departed Port Jackson on 27 May, accompanied by the destroyer HMAS *Duchess*. Destroyer escort HMAS *Parramatta* joined on 2 June. The group rendezvoused on 4 June with HMAS *Melbourne*, *Vampire*, and the supply ship *Supply*, southeast of the Philippines, to replenish.[3]

TRANSPORT/SUPPORT OF AUSTRALIAN TROOPS

The task of transporting, supplying and maintaining Australian forces in South Vietnam was shared by the Royal Australian Air Force, civilian aircraft (mainly Qantas), ships from the Australian National Line (ANL), and the Royal Australian Navy. The bulk of this work was shouldered by the RAN, and the ship that carried out the majority of such duties was the former carrier, HMAS *Sydney*.[4]

HMAS *Sydney*, which would become known as the "Vung Tau Ferry," had been launched at Devonport Dockyard, United Kingdom, on 30 September 1944. Unfinished at the end of World War II, work on the *Majestic*-class light fleet aircraft carrier ceased. After the Australian government purchased the ship, remaining work on her was completed, and she was handed over to the Royal Australian Navy on 16 December 1948 and named *Sydney*.[5]

Sydney served in Korean Waters from September 1951 to January 1952, during which her aircraft flew 2,366 sorties. She sailed for another period in Korea on 19 October 1953, and returned to Australia in June 1954 for a refit. Following the war, *Sydney* became a fleet training ship in April 1955. Paid off on 30 May 1958, she sat idle in Sydney Harbour (berthed at Athol Bight near Bradleys Head) until recommissioned as a troop transport on 7 March 1962. She might have remained laid up in "mothballs," had not the Australian Army "top brass" pointed out that in the event of a limited war, or need to counter an insurgency in

Southeast Asia, the former aircraft carrier would be of enormous value in the movement of troops, vehicles, ammunition, and other stores and equipment.[6]

Photo 4-2

1RAR formed up on the flight deck of HMAS *Sydney*, en route to Vietnam. Courtesy of Sea Power Centre – Australia

Nearing Vietnam, *Melbourne* and *Vampire* detached in early morning on 8 June, leaving *Duchess* and *Parramatta* to escort the troop transport to Vung Tau. *Sydney* and her escorts anchored there a short time later, and cargo unloading began immediately. It continued until 11 June, whereupon the ships proceeded to Singapore, en route return passage to Australia. *Sydney* made another voyage to Vietnam in September 1965. She was initially escorted by HMAS *Anzac* and *Melbourne* before

HMAS *Duchess* and *Vendetta* took over these duties for the final transit between Manus Island (Papua New Guinea) and Vung Tau.⁷

Photo 4-3

DeLong Pier at Vung Tau with the RAN bulk carrier HMAS *Jeparit* alongside. Australian War Memorial photograph P03051.002

On 8 March 1966, Australian Prime Minister Harold Holt committed additional troops to the war. The 1st Australian Task Force (1 ATF) was to be based at Nui Dat, Phuoc Tuy Province. *Sydney* made two voyages in May and June, carrying 5RAR and 6RAR to Vietnam. *Melbourne* again served as an escort. With Australian ground forces now well established in Vietnam, *Sydney* began a regular pattern of disembarking a battalion at Vung Tau, and back loading one rotating out of combat, for return passage to Australia. After her escort was reduced to a single ship, a detachment of four Wessex MK 31A helicopters from 725 Squadron were embarked in April 1967 to provide *Sydney* additional anti-submarine protection. The Wessex flight (group of aircraft) was replaced by a similar flight, from 817 squadron, for five voyages between December 1967 and December 1968.⁸

Sydney made three voyages to Vietnam in 1967, four in 1968, three in 1969 and two in 1970. Her busiest year came in 1971 when she deployed to Vietnam six times. Overcrowding existed whenever Army personnel were embarked, requiring many junior sailors to sleep in 'A' Hangar to free up mess deck accommodation for the troops. Many of the young sailors assigned to *Sydney* were 16-year old junior recruits getting their first taste of life at sea.⁹

Photo 4-4

HMAS *Sydney* in her role as the "Vung Tau Ferry," circa 1968.
Courtesy of Sea Power Centre – Australia

SHIPBOARD DEFENSIVE MEASURES AT VUNG TAU

While lying at anchor at Vung Tau, HMAS *Sydney* and other Australian ships employed a layered approach of self-protective measures against enemy swimmer-sappers, who endeavored to employ explosive devices against ships at anchor or berthed alongside a pier. Safeguards included the posting of armed, upper-deck sentries, additional lookouts, waterborne patrols, and inspections of the hull and anchor cables by ship's diving teams. Underwater scare charges (concussion grenades) were also used as a deterrent against enemy swimmers.[10]

Photo 4-5

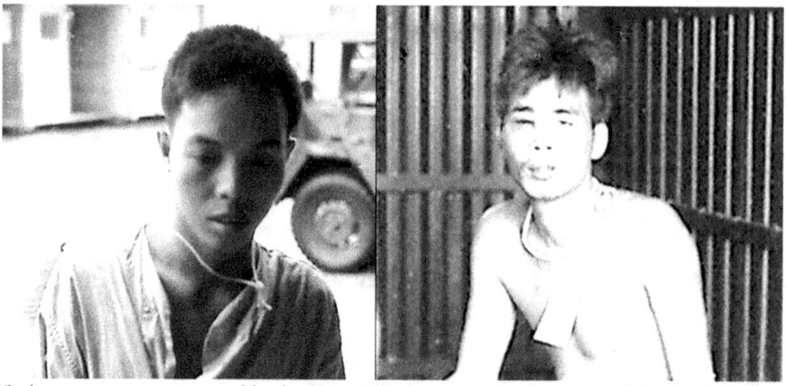

Swimmer-sappers captured in the Vung Tau port area on the night of 22 May 1969, following an unsuccessful attempt to blow up the motor vessel *Heredia*—which, with almost 8,000 tons of explosives aboard, was functioning as an ammunition ship. The enemy-emplaced charges were found and rendered harmless by members of the Royal Australian Navy's Clearance Diving Team 3, based at Vung Tau.
Courtesy of Hector Donohue

WITHDRAWAL FROM VIETNAM

By late 1971, the withdrawal of Australian forces from Vietnam had begun. *Sydney*'s mission then shifted from rotating infantry battalions, to bringing them home. At Vung Tau on 8 December 1971, she embarked 4RAR (the final battalion), the 104 Field Battery, and No. 9 Squadron RAAF with its sixteen Iroquois helicopters. Australia's combat role in South Vietnam ceased in March 1972, when *Sydney* returned home the last combat elements.[11]

5

HMAS *Boonaroo* and *Jeparit*

> *It must have been one of the shortest and least ceremonious commissionings ever held, consisting as it did simply of reading the commissioning warrant and hoisting and lowering the ensign, all in the dark.*
>
> —Comdr. Patrick R. Burnett, RAN, remarking on the commissioning of the former Australian merchant vessel MV *Boonaroo* into the Royal Australian Navy as HMAS *Boonaroo*, after her former merchant crew refused to sail the ship with a cargo of ordnance to Vietnam.[1]

Photo 5-1

Postcard of MV *Boonaroo*, location and date unknown.

In 1966, the Australian Department of Shipping and Transport chartered the 391-foot, diesel-powered cargo ship MV *Boonaroo* to carry supplies to Vietnam in support of Australian forces. She made one round-trip voyage, departing on 17 May and returning on 8 July 1966.[2]

In February 1967, the Seaman's Union refused to sail the Australian National Line ship to Vietnam with a cargo of Royal Australian Air Force ordnance. Consequently, she was taken up from trade, and commissioned HMAS *Boonaroo* on 1 March 1967. By coincidence, she was the first ship of the Royal Australian Navy to be commissioned under the Australian White Ensign. On that day, the new flag was proudly hoisted aloft, in lieu of the British White Ensign (previously flown aboard RAN vessels), with one unique to Australia.[3]

With commissioning came a crew change aboard ship—eliminating the problem with Seamen's Union members. Comdr. Patrick R. Burnett, RAN, relieved the ship's master, Captain P. Grimanes and, except for two engineer officers with Naval Reserve commissions, the entire crew was replaced with Royal Australian Navy officers and men.[4]

Australian White Ensign (1967-present)   British White Ensign (1911-1967)

Boonaroo left Melbourne at 0500 the next morning for the explosives berth at Point Wilson (on the northern shore of Corio Bay, Victoria) to load cargo. This consisted mainly of 500 and 1,000lb bombs, but there were vehicles and other items also destined for the air base at Phan Rang (home of No. 2 Squadron, RAAF), about twenty-five miles south of Cam Ranh Bay. Loading was completed by 10 March, and *Boonaroo* sailed for Cairns. A stop at a base on the shore of Trinity Bay, in Queensland, would provide opportunity to top off fuel before departing Australia.[5]

With a modest top speed of 11½ knots, it took *Boonaroo* seven days to reach Cairns. Arriving on the morning of 17 March, she left that same day after fueling. The ship cleared Torres Strait, separating Australia and Papau New Guinea, on the 19th, and continued her passage west of West Irian and east of Celebes and Borneo. She arrived off Cam Ranh Bay at 0700 on 28 March, with her crew at Action Stations (General Quarters), not knowing quite what to expect. Cam Ranh lay about 200 miles north of Vung Tau, the normal Australian logistic site in Vietnam.[6]

Map 5-1

Australia

During *Boonaroo*'s five-day stay for unloading, Commander Burnett was given a jeep tour of the base, which covered an extensive area of sandhills. The local front line was in the coastal ranges several miles to the west, held by a South Korean battalion, where the sound of firing and air attacks could occasionally be heard. The only threats to the base were infrequent hit-and-run grenade and mine attacks.[7]

Leaving Cam Ranh Bay on 2 April, *Boonaroo* proceeded south, and anchored in the open sea off Vung Tau, for the transfer of stores and mail by RAAF helicopters. While there, she exchanged greetings with MV *Jeparit*, the second Australian National Line (ANL) vessel to be requisitioned to take war supplies to Vietnam. *Jeparit* was unloading at Vung Tau, manned by merchant mariners with a RAN detachment replacing members of the Seamen's Union. *Boonaroo* arrived back at Melbourne on Friday, 5 May, berthing at No 11 North Wharf in the Yarra River, late that evening.[8]

The weekend was spent disembarking naval stores and the ship's company to HMAS Lonsdale, a naval depot located at Beach Street, in Port Melbourne. In mid-afternoon on Monday, 8 May, a brief decommissioning ceremony was held at which the ship was officially handed back to the ANL. So ended the sixty-nine-day naval service of HMAS *Boonaroo*—the first, and only occasion on which the RAN has commissioned and operated a merchant ship in peacetime because of a union labour dispute.[9]

BALLAD OF THE *BOONAROO*
It was the good ship *Boonaroo*
That sailed the Aussie main
Like many other freighters do,
Which, sailed by their civvie crew,
Drop cargo off at Wallaroo
And home return again.
But then one day the call was heard
To Vietnam we go!
The Seaman's Union quickly stirred;
A meeting soon sent out the word,
They promptly gave the trip the bird
And loudly answered 'No!'
And so the *Boonaroo* was manned
By sailor boys in blue;
An operation quickly planned
By the Top Brass throughout the land;
The Navy lent a helping hand
As it will always do.
Now HMAS *Boonaroo*
Sails o'er the seven seas;
Sweethearts and wives, stay always true,
Manning the proper thing will do
And post us back again to you
To live a life of ease.
(author unknown)[10]

MOTOR VESSEL *JEPARIT* SAILS WITH MIXED CREW

HMAS *Jeparit* ship's crest   HMAS *Boonaroo* ship's crest

The motor vessel *Jeparit* was chartered by the Department of Shipping and Transport in June 1966, to carry equipment and supplies from Australia to South Vietnam for the 1st Australian Task Force. In February 1967, the Seamen's Union refused to man her, as it had *Boonaroo*. However, in the case of *Jeparit*, existing crew agreeable to continuing to serve in the ship were supplemented by a Royal Australian Navy detachment. Sailing under the Red Ensign of the Merchant Marine, with a mixed crew of twenty merchant mariners and a RAN detachment of one officer (Lt. Robert Winter) and seventeen sailors, *Jeparit* made twenty-one voyages to Vietnam and back.[11]

Australian Red Ensign flown by Merchant Marine vessels.

In November 1969, the Waterside Workers Federation refused to load or unload *Jeparit* in Sydney, associated with anti-war sentiment. This action prompted the commissioning of *Jeparit* on 11 December as one of Her Majesty's Australian Ships, under the command of Comdr. Richard Bourke, RAN. The following day the ship's master, Captain A. A. C. Philip, was given a commission as a commander in the Naval Volunteer Reserve and assumed command from Bourke. HMAS *Jeparit* made a further seventeen voyages under the Australian White Ensign. In total, she carried 175,000 deadweight tons of cargo to Vietnam before her return to ANL control in March 1972.[12]

Photo 5-2

Jeparit with the Sydney Harbour bridge in the background.
Courtesy of Sea Power Centre – Australia

Photo 5-3

RAN bulk carrier *Jeparit* alongside De Long Pier at Vung Tau.
Courtesy of Sea Power Centre – Australia

FORMER GERMAN OFFICER ABOARD *JEPARIT*

> *If this operational record is impressive, then in economic terms, the figures are even more so, as this outstanding performance was achieved by little more than 3,000 officers and men, in nine second-hand freighters, armed with third-hand weapons, the total cost of which, both in terms of purchase price and the cost of fitting out, represented barely 1% of the cost of the battleship* Bismarck*!*
>
> —Reference to the nine former German freighters converted to auxiliary cruisers (commerce raiders) during World War II to prey on Allied merchant shipping. These wolves in sheep's clothing accounted for 141 ships sunk or captured in the war.[13]

Many of the MV/HMAS *Jeparit* merchant mariners had interesting backgrounds, but none more so than Chief Officer Alfons Schmitt. During World War II, Schmitt had served briefly aboard the 13,580-ton German tanker/supply ship *Altmark*. Commanded by Capt. Heinrich Dau, she had, for a time, been the supply ship for the pocket battleship *Admiral Graf Spee* off the east coast of South America. After suffering extensive damage at the hands of three allied cruisers—HMS *Exeter*, HMS *Ajax*, and HMNZS *Achilles*—in the Battle of the River Plate, Captain Dau had scuttled *Graf Spee* on 17 December 1939 to avoid the possibility of her capture.[14]

After leaving *Altmark*, Schmitt continued his war service in the German commerce raider HK *Pinguin*. She was one of nine nondescript, converted German freighters (disguised as "clapped out" merchantmen of other nations) that went to sea and fought as true warships. Armed with hidden guns, torpedo tubes, and mines, and carrying a Heinkel seaplane to scout for prey, *Pinguin* left Sorgulenjord, Norway, on 22 June 1940, bound for the Antarctic and Indian Ocean. Her commanding officer, Kapitan zur See Ernst-Felix Kruder, had placed the ex-freighter *Kandelfels* into commission on 6 February with a wartime complement of 17 officers, five prize officers, and 398 petty officers and men.[15]

Kruder's brilliant record of thirty-two ships captured or sunk came to an end on the morning of 8 May 1941, near the entrance to the Persian Gulf. She was no match for the 8-inch guns of the Royal Navy heavy cruiser HMS *Cornwall*, and was sunk, taking with her 341 Germans and 214 prisoners from vessels she had captured. *Cornwall* (Capt. Percival C. W. Manwaring, RN) was able to rescue 24 British and Indian prisoners and 60 German sailors—amongst them, Alfons Schmitt.[16]

Photo 5-4

The German raider HK *Pinguin* in the Indian Ocean, 1941.
Australian War Memorial photograph P02018.032

Schmitt, with other survivors, was taken to Australia as a prisoner of war. He declined repatriation to Germany at war's end, remained in Australia, and joined her merchant navy.[17]

UNITED STATES' SEA-SUPPLIED LOGISTICS

The U.S. counterpart to the Royal Australian Navy's logistics/troop ships—HMAS *Sydney*, *Boonaroo*, and *Jeparit*—were merchant vessels of the Military Sea Transportation Service (MSTS). Although there was some interface between the two and overlap of responsibilities, generally MSTS ships supplied U.S. Army and Air Force commands in-country, while Seventh Fleet Service Force ships provided logistics to Navy ships offshore—aircraft carriers at Yankee and Dixie Stations, surface combatants on the gunline, and amphibious forces making landings along the South Vietnamese coast—and to Navy and Marine Corps units in-country.

The supply of military forces in-country by sea, whether they be Army, Air Force, Navy, or Marine Corps, required port development, and associated surveys of coastal and harbor waters for the production of nautical charts needed for ships to safely navigate the approaches to, and within the harbors where ports were located. This requirement was met by six Seventh Fleet survey ships. None were purpose-built (being former attack cargo ships, minesweepers, a seaplane tender, and a tug), and all were old, but they got the job done.

6

Survey Ship Operations

The first sight of Vietnamese coastal waters was a choppy, muddy brown—the mouth of the Bassac River, a part of the vast Mekong Delta. Here we made a brief exploratory survey. The Viet Cong bid us our first welcome to Viet Nam by firing at soundboat Eight—*no one was hit, and "Tiger Leader" (Mr. Paul) and crew warmly returned the welcome.*

—Description of the initial work by survey ship USS *Maury* (AGS-16) during her deployment with USS *Serrano* (AGS-24) to Vietnam in 1965-1966, to survey and develop charts for coastal waters. Encounters with the enemy resulted in *Maury* being awarded a combat action ribbon for 11 December 1965 and, during a subsequent deployment, *Serrano* for 3-4 March 1967.[1]

Photo 6-1

Survey ship USS *Maury* at anchor, with USS *Serrano* and a sound boat alongside.
USS *Maury* (AGS-16) and *Serrano* (AGS-24) Vietnam Survey 1965-1966 cruise book

It became evident by November 1965 (several months after the 9th Marine Expeditionary Brigade landed at Da Nang on 8 March), that coastal charts for South Vietnam, based on Japanese hydrography from World War II, were unreliable, particularly in the river deltas. Over the next three years, the Naval Oceanographic Office completed comprehensive geodetic, coastal, and harbor surveys of the complex coastline using a series of survey vessels, none of them purpose-built.[2]

Finding itself with too many of one type of ship, and not enough of another, the U.S. Navy is apt to convert the former to make up shortfalls in the latter. Such was the case with the six survey ships that served in Vietnam. *Maury* and *Tanner* were former attack cargo ships; *Sheldrake* and *Towhee* were ex-minesweepers. *Rehoboth* began her naval service as a seaplane tender, and *Serrano* as a fleet tug. Some characteristics of these ships are provided in the table, as well as their original commissioning date, followed by the date they were recommissioned as survey ships, and their ultimate decommissioning date.

Oceanography Survey Ships that Served in Vietnam

Survey Ship	Length (feet)/ Displ. (tons)	Commissioned Decommissioned
Maury (AGS-16) ex-USS *Renate* (AKA-36)	426/7,080 (full load)	28 Feb 45/12 Jul 46 19 Dec 69
Rehoboth (AGS-50) ex-USS *Rehoboth* (AVP-50)	310/2,800 (full load)	23 Feb 44/2 Sep 48 15 Apr 70
Serrano (AGS-24) ex-USS *Serrano* (ATF-112)	205/1,675	22 Sep 44/30 Jun 60 2 Jan 70
Sheldrake (AGS-19) ex-USS *Sheldrake* (AM-62)	221/890	14 Oct 42/14 Apr 52 1 Aug 68
Tanner (AGS-15) ex-USS *Pamina* (AKA-34)	426/7,080 (full load)	10 Feb 44/15 May 46 1 Aug 69
Towhee (AGS-28) ex-USS *Towhee* (AM-388)	221/890	18 May 45/1 Apr 64 30 Apr 69

Survey ships *Maury* (Capt. Robert Francis Reilly, USN) and *Serrano* were dispatched to Vietnam in late 1965. They began their first major survey of the cruise at Cam Ranh Bay in mid-December. The large natural harbor on the coast of central Vietnam was largely undeveloped; surrounded by mountains and high sand dunes, it offered calm waters in its natural state. However, this condition was rapidly changing with the development of a large logistics port. When *Maury* arrived (she was joined later by *Serrano*, journeying from Subic Bay), there were more than 12,000 U.S. Army, U.S. Air Force, and Korean Marine forces living in tent cities on the sand dunes. Additionally, more than twenty cargo ships were waiting in the harbor to unload materiel for the support of American combat troops in-country.[3]

The job of the survey ships and their 50-foot wooden-hulled soundboats was to survey the bay and its environs, to provide charts for navigation, and information for further development of port facilities. This effort began with geodetic work, and setting up the first of several base camps. Personnel at the camps provided electronic "fixes" from RAYDIST equipment sited ashore, to correlate/pinpoint the location of the ships' fathometer soundings. RAYDIST (an acronym of radio and distance) was a radio system for medium-range precision surveying in which the phases of two continuous-wave signals were compared. In a few days, soundboats began "running lines" in shallow waters nearer the shore.[4]

Photo 6-2

At left, *Maury* personnel construct a visual geodetic signal; at right, *Maury* offloads a soundboat to begin survey work.
USS *Maury* (AGS-16) and *Serrano* (AGS-24) Vietnam Survey 1965-1966 cruise book, and USS *Serrano* (AGS-24) and *Maury* (AGS-16) 1964-1965 cruise book, respectively

The initial survey work was briefly interrupted on 19 December, when *Maury* was ordered to join a search and rescue operation for the crew of the *Impala*. The Panamanian merchant ship had struck a shoal two miles south of Cape Varella, at approximately 0520 that morning, and sank. At 0700, the fast combat support ship *Sacramento* (AOE-1) discovered a lifeboat with seven survivors and a deceased crewman on board. One of those saved, the first mate, reported that all twenty-nine men of the crew had cleared the ship with life jackets, but were swept out to sea by strong currents. *Sacramento* assumed duties as on scene commander, and began searching for other possible survivors.⁵

At 1030, the destroyer *Henderson* (DD-785), ammunition ship *Firedrake* (AE-14), ocean minesweeper *Gallant* (MSO-489), and *Maury* were directed to join *Sacramento* and assist in the search. P-3 patrol aircraft and helicopters were already on the scene. By late afternoon the following day, twelve bodies had been recovered. When the *Sacramento* and *Firedrake* were detached to carry out prior commitments, *Maury* assumed duties as on scene commander. Searching continued until nightfall. *Gallant* and *Henderson* were then detached as the possibility of sighting any additional survivors or bodies in darkness was considered minimal. *Maury* remained in the area until early afternoon on the 21st, but found no additional survivors.⁶

Photo 6-3

USS *Maury* (AGS-16), location and date unknown.
USS *Maury* (AGS-16) and *Serrano* (AGS-24) Vietnam Survey 1965-1966 cruise book

The *Sacramento*, *Firedrake*, and *Maury* were Service Force ships, deployed to Vietnam in support of the U.S. Navy Seventh Fleet ships engaged in combat duties. These included cruisers and destroyers on the gunline off South Vietnam, amphibious ships engaged in landings all along the coast, salvage ships, and others. Among the others was the 172-foot, wooden-hulled *Gallant*.

Photo 6-4

USS *Gallant* operating with Mine Division 73. Top to bottom: *Illusive* (MSO-448), *Conquest* (MSO-488), *Esteem* (MSO-438), *Gallant* (MSO-489), and *Pledge* (MSO-492). U.S. Navy photo from the August 1965 edition of *All Hands* magazine

She and the other units of Mine Division 73 were engaged in Operation MARKET TIME, a massive combined U.S. Navy and South Vietnamese Navy effort to stop Viet Cong infiltration of weapons and supplies by sea into South Vietnam. In 1965, U.S. Navy ships were formally directed to assist the South Vietnamese Navy in its coastal surveillance and anti-infiltration efforts. Throughout the war, divisions of minesweepers would depart the West Coast of the United States every four months, headed west to relieve the division on station that was rotating home, often proceeding with lights darkened and emissions control set (the restricted use of radar and communications equipment to avoid detection) preparing for combat duty.[7]

RESUMPTION OF SURVEY OPERATIONS

Maury returned to Cam Ranh Bay and by Christmas Day, the survey was well under way. The day after Christmas, *Maury* and *Serrano* stood out of the bay, bound for Cebu City, on the island of Cebu in the central Philippines, for crew liberty. Founded in the sixteenth century by Ferdinand Magellan, the "Great Navigator," Cebu had become the first Spanish capital of their Philippine colony. Among the sights around Cebu was a memorial that included a splinter from a cross placed by Magellan. He was later killed, on 27 April 1521, by a poison arrow fired by Chief Lapu-Lapu. His death occurred during a skirmish on the island of Mactan, while Magellan and his crew were assisting a local king they had allied with after landing on Cebu weeks earlier.[8]

Photo 6-5

Memorial containing a splinter from Magellan's cross.
USS *Maury* (AGS-16) and *Serrano* (AGS-24) Vietnam Survey 1965-1966 cruise book

The ships arrived back at Cam Ranh Bay from Cebu, to find that two soundboats left behind had completed much of the survey work inside the bay. While nearshore efforts continued, beach camps were established on offshore islands to extend the survey to the adjoining coast, and at Nha Trang to the north and Phan Rang to the south. This action involved a considerable dispersal of survey and support units.[9]

Photo 6-6

Operation MARKET TIME base site, Cam Ranh Bay, November 1965.
Naval History and Heritage Command photograph #NH 74203

Photo 6-7

RAYDIST transmitting station on Hon Ngoi Island.
USS *Maury* (AGS-16) and *Serrano* (AGS-24) Vietnam Survey 1965-1966 cruise book

SIDE TRIP TO VUNG TAU

A few weeks into 1966, *Maury* left for a brief period to survey an area at Vung Tau, near the mouth of the Saigon River. This old port city and popular seaside resort (known in the French colonial days as Cap. St. Jacques), seemed comparatively unaffected by the hostilities—except for the presence of American servicemen. There was a near-holiday atmosphere at the beach, a wide sandy strip lined with stucco "bar-restaurants" offering Vietnamese "Biere Larue" and "33." Although these beverages didn't match the brew back home, they were wet, cold (if you were lucky), and had the same effect.[10]

LOGISTICS SUPPORT AT CAM RANH BAY

During her time spent at Cam Ranh Bay, *Maury* replenished from the *Sacramento* and the combat stores ship *Mars* (AFS-1). In addition to receiving fuel, helicopters from the two service ships delivered to *Maury* a steady supply of fresh fruits and vegetables via the newly developed VertRep (vertical replenishment) method.[11]

Photo 6-8

Vietnamese command junk on patrol.
USS *Maury* (AGS-16) and *Serrano* (AGS-24) Vietnam Survey 1965-1966 cruise book

Binh Ba Island (at the entrance to Cam Ranh Bay) was the site of a Vietnamese Navy "junk force" base, from which its small patrol craft monitored coastal junk traffic which might be carrying supplies to the Viet Cong. Survey ship sailors and those of the junk force frequently visited one another for support and assistance. As a result of this interaction, *Maury*'s officers and crew were made honorary "junkies" by the South Vietnamese commanding officer, who authorized them to wear the black beret with silver junk force emblem while in Vietnam.[12]

Photo 6-9

Members of the Vietnamese "junk force."
USS *Maury* (AGS-16) and *Serrano* (AGS-24) Vietnam Survey 1965-1966 cruise book

By mid-February 1966, the Cam Ranh survey was completed. The soundboats and beach camps were loaded aboard, and the survey ships proceeded south to the mouth of the Bassac River. A survey there was rapidly completed, with the soundboats working around the clock. At completion, the ships departed for Subic Bay. *Serrano* left Subic after a short stay to return to work, conducting surveys around Con Son Island (the largest island in the Con Dao Archipelago), and the Royal Bishop Banks in the South China Sea.[13]

Map 6-1

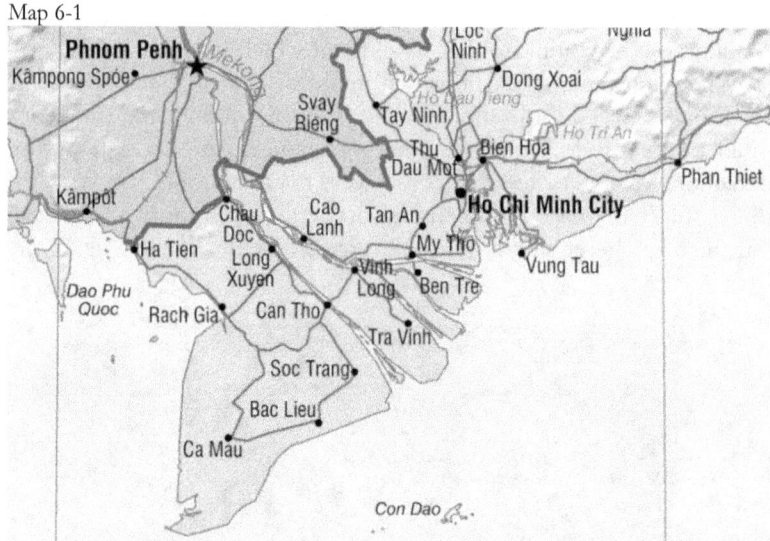

Con Dao Archipelago. Ho Chi Minh City was formerly known as Saigon.

While *Serrano* carried out this work, and found time to visit Vung Tau, *Maury*'s crew enjoyed a full two weeks in Subic. For some, this meant relaxation in the mountains at Baguio City in Northern Luzon; for others it was action in the city of Manila, the capital of the Philippines.[14]

SURVEY OPERATIONS AT CHU LAI

Photo 6-10

Chu Lai Peninsula, Vietnam, January 1966.
Naval History and Heritage Command photograph #NH 74192

In mid-March, the ships rejoined to conduct survey operations at Chu Lai. A major U.S. Marine Corps-held area, it lay well up the South Vietnamese coast, centered on a dusty bluff fifty-seven miles south of Da Nang. The marked contrast between it and the relatively quiet and secure Cam Ranh Bay became apparent shortly after their arrival, with the Marines embarking on a major operation against Viet Cong forces in the area. As *Maury* and *Serrano* made initial preparations to survey the coastal waters and small harbor area, sailors witnessed the continuing coming and going of Marine helicopters and, by night, mortar bursts and flares in the distance.[15]

Beach camps were set up with the Marines providing helicopter transport to the sites, and additional security. The value of the latter service was validated when an unsuccessful attempt was made by the enemy to destroy the transmitting tower at "Red Station" with hand grenades.[16]

SOUNDBOAT 7 HIT BY ENEMY FIRE

Photo 6-11

Machine gunner and lookout/loader aboard Soundboat 7 monitor the shoreline, for any indication of enemy forces, during survey operations in waters off Chu Lai. USS *Maury* (AGS-16) and *Serrano* (AGS-24) Vietnam Survey 1965-1966 cruise book

48   Chapter 6

When operations began, all six units—*Serrano*, the four soundboats, and for the first time, *Maury* as well—joined in running survey lines. While working near Chu Lai, Soundboat 7 was taken under fire by automatic weapons from the beach, about 150 yards away. Crewmembers quickly returned fire with small arms. As the coxswain swung the boat to retire seaward, a second burst cut across the bow at deckhouse level. Several bullets struck the craft; one passed through a window, just missing a sailor operating the fathometer. The boat's .50-caliber machine gun opened fire—silencing the enemy. The officer in charge, Lt. (jg) Don Puccini, was credited with his crew's rapid actions in manning their stations and returning fire; thereby minimizing damage and averting casualties. Seaman Paul Brophy was the machine gunner.[17]

Following a visit by the two ships to Bangkok, Thailand, and their return to Chu Lai on 20 April, beach camps were reestablished. *Serrano* and two soundboats then set out to complete the survey. Five days later, *Maury* (with two soundboats) proceeded north to Da Nang, to begin surveying a new area of the coast near Hue. Following establishment of a RAYDIST navigational net, a storm from the northeast brought rising winds, and the seas became choppier each day. Despite these unfavorable conditions, all work was completed by 4 May, and *Maury* returned to Chu Lai.[18]

VISIT TO HONG KONG

> *Wan Chai, just east of the central district, is a place well-designed to part a sailor from his money. Aside from the numbers of familiar bars, nightspots and restaurants, a main attraction was the nearby China Fleet Club's large Navy contract arcade, with displays by dealers in all kinds of Hong Kong merchandise at reduced prices to servicemen. With such prices, averaging half those back in the states for similar items, it was a great temptation to "go broke saving money." The best and most popular bargains were in high-fidelity equipment, cameras, watches, jewelry, souvenirs, and especially the famous Hong Kong clothes. Suits, shirts, and shoes made from fine English and local materials are custom tailored in three to four days.*
>
> —Description from USS *Maury* and USS *Serrano*'s 1965-66 cruise book, of the many bargains available in Hong Kong in 1966.[19]

All survey work at Chu Lai was completed on 10 May, and *Maury* and *Serrano* headed back to Subic Bay. Following a stay of several days, during which they refueled and re-provisioned, both ships got under way for Hong Kong. The two-day passage across the South China Sea was rough, as they were on the periphery of Typhoon IRMA. Happily, the weather was clear upon arrival at Hong Kong on the 18th, and remained so during their five-day visit, while the typhoon raged back in the Philippines.[20]

Photo 6-12

Hong Kong and Kowloon by night.
USS *Maury* (AGS-16) and *Serrano* (AGS-24) Vietnam Survey 1965-1966 cruise book

An impressive sight greeted the ships upon entering the harbor. On both the Hong Kong and Kowloon sides, innumerable high-rise buildings appeared to cover all usable space along the coastline and up the steep hillsides. (Kowloon, on the mainland across Victoria Harbour, was an urban area in Hong Kong comprising the Kowloon Peninsula.) The ships moored in the harbor, just off the central district of Hong Kong. Getting ashore was easy. The famous Star Ferries to Kowloon passed close by on one side; and "walla-walla" boats on the other, which made the short jaunt to the fleet landing in the Wan Chai District at Fenwick Pier.[21]

Shopping, eating at the many fine restaurants, and enjoying the nightlife of Hong Kong left little time for sightseeing. Those who took the time to do so, experienced a spectacular view from the top of "the Peak" on Hong Kong Island, reached by a steep tram. From this (high) vantage point, the panoramic view included typhoon shelters crowded with junks; and the floating restaurants on the other side of the island from the central district.[22]

MAURY'S DEPLOYMENT EXTENDED

The Hong Kong visit was to have been the last major port call before return to Pearl Harbor at deployment's end. However, while en route to Hong Kong, *Maury* had learned that she was to be sent to another survey area. Accordingly, upon leaving the British Crown Colony, the survey ships parted company. *Serrano* was bound for Hawaii, and *Maury*, to Dam An Hai—a shallow lagoon between Da Nang and Hue.[23]

The lagoon proved to be very shallow indeed, requiring most of the survey to be carried out by *Maury*'s small twelve-foot wherry, dubbed "soundboat *9*." The necessary work required only a brief stay before *Maury* was able to set a course for Pearl Harbor. Shortly before her departure, she was joined by the *Glinta*, a chartered merchant vessel, aboard which was a civilian oceanography team.[24]

During the cruise, soundboat surveys had been more extensive, and important than ever before. Five thousand miles of survey lines were run in Vietnamese waters, from the Bassac River mouth on the Mekong Delta, to Hue in the north. While about this work, "Adrifts Seven and Eight" (call signs for soundboats) had come under enemy fire, promptly returned fire, and safely escaped. Adrift One had provided communications support for the 101st Airborne Division during a landing north of Phan Rang, and Adrift Two had tallied more survey miles than any other boat, while participating in all survey areas except Hue.[25]

CONCURRENT WORK BY *REHOBOTH*

During *Maury* and *Serrano*'s deployment, survey ship *Rehoboth* (AGS-50) was similarly employed. She commenced survey operations in the South China Sea in November 1965, and during December conducted a hydrographic survey of the South Vietnamese coast from the Mekong Delta to Cape Padaran (located midway between Phan Thiet to the south, and Cam Ranh Bay to the north). After completing survey operations in the South China Sea in February, she set a course for San Francisco, arriving on 23 March 1966.[26]

7

Maury and *Serrano* Return to Vietnam

> *A 52-foot, 32-ton soundboat from the hydrographic survey ship USS* Maury *AGS-16 was taken under hostile fire near the Co Chien River on 13 May [1967] while conducting close-inshore surveying. Struck below the waterline by the first of several 57mm recoilless rifle shots, the soundboat crew immediately returned fire with .50 caliber, as did the ... escort WPB [Coast Guard cutter Point]* Kennedy *with both .50 cal. and 81mm high explosives.... [The] officer in charge, LTJG Phil Lamberson, ... ordered his soundboat to clear the area to seaward and the crewmen began immediate damage control efforts.*
>
> —Excerpt from an article, "Soundboat Hit by VC!!," in the *Pathfinder* newspaper, Vol. IV No. 1, USS *Maury* (AGS16) May 1967.[1]

Maury and *Serrano* returned to Vietnam in 1967, engaged for the first nine months of the year in surveying various mouths of the Mekong River and Nha Trang area. Fourteen field charts were constructed aboard *Maury*, in four colors for greater ease in use. For this important accomplishment, the secretary of the Navy awarded the ships a Meritorious Unit Commendation.[2]

Maury and *Serrano*'s charter was to provide field charts to place new information in the hands of ship navigators at the earliest possible time. Chart production required the many talents and abilities of their crewmembers. Surveyors surveyed and computed geodetic positions of reference to pinpoint soundings taken by the soundboats. Draftsmen used this data to prepare boatsheets, and then smoothsheets (filled with these soundings), which were then photographed.[3]

From the photographs, soundings were selected to represent the depths in areas on the chart original. The chart original, which also had the coastline and parallels of latitude and longitude drawn on it, was then photographed by the printer to produce lithographic plates from which a chart was reproduced for use by the fleet. Seabees (Naval

52   Chapter 7

Construction Battalion personnel), printers, photographers, Marines, and skilled civilians aboard the ships, worked together to produce an essential contribution to the Vietnam War effort.[4]

Photo 7-1

Serrano, *Maury*'s "running mate," moored alongside the larger of the two survey ships. USS *Maury* (AGS-16) Kwajalein-Vietnam Survey 1966-1967 cruise book

Maury's cruise began on 4 January 1967 when, as bands played and dependents wept on the pier, she passed out through the Pearl Harbor channel, Vietnam bound. *Serrano* had earlier sailed west from Pearl on New Year's Day. *Serrano* arrived at Vung Tau by month's end, then proceeded north to begin a Da Nang-Hue coastal survey. In March, *Serrano* operated out of the Song Cua Dai Junk Base where her medical personnel provided care for the residents of the area. She replenished in the Philippines at the end of April, then resumed work on the Da Nang-Hue survey which ended on 18 July. Following completion of her last Vietnam assignment, *Serrano* set a course for Pearl Harbor, arriving on 29 September.[5]

The remainder of this chapter is devoted to the activities of *Maury*. Being significantly larger than *Serrano* (a former fleet tug), she had much greater capabilities resident aboard her, including personnel and facilities necessary for chart production. (Having such also facilitated production of expansive cruise books, which is why many of the photos herein are from her cruise books.)

SOUNDBOAT *1* NEARLY LOST IN MEKONG DELTA

Photo 7-2

Survey ship USS *Maury*'s (AGS-16) Soundboat *8*, a sister to Soundboat *1*.
USS *Maury* (AGS-16) Kwajalein-Vietnam Survey 1966-1967 cruise book

Upon her arrival at Vung Tau on 13 March 1967, *Maury* began boat survey operations in the Mekong Delta. From anchorage at Vung Tau, she provided support for her soundboats, which were working the lower reaches of the Delta in conducting an inshore hydrographic survey. As indicated in the material at chapter's head, Soundboat *1* was hit by 57mm recoilless rifle fire on 13 May—two months into this effort.[6]

The term "lower reaches" refers to the area of a river nearest the sea into which it flows, and the boat was near the Co Chien River, when she was nearly sunk by enemy fire. As Lieutenant (junior grade) Lamberson turned seaward to open the coast, crewmembers worked furiously to plug a 15 by 20-inch hole at the waterline, which (along with 3-inch fragment holes below it) was allowing water to pour into the boat at a rate of 400-500 gallons per minute. This influx was greater than the capacity of the boat's pumps. Fortunately, the 82-foot Coast Guard cutter *Point Kennedy* was nearby to provide assistance.[7]

Photo 7-3

USCG cutter *Point Kennedy* (WPB-82320), 6 January 1967.
National Archives photograph #K-35540

The cutter was already on the scene, having been assigned to provide gunfire support for Soundboat *1* during survey operations. After suppressing the shore fire, she took the boat alongside, and assisted damage control and dewatering with two electric submersible pumps, additional shoring material, and a salvage pump. Meanwhile, *Maury* proceeded at maximum speed toward the sinking soundboat from her position twenty-five miles northward, and the tank landing ship *Harnett County* (positioned at the river mouth to support patrol boats and assault helicopters) dispatched another pump.[8]

Photo 7-4

Navy UH-1B Iroquois ("Seawolf") helicopters based aboard USS *Harnett County* (LST-821), take off on a strike mission against the Viet Cong, September 1969. National Archives photograph #K-58279

This added capability, in conjunction with the efforts of the boat crew and the Coastguardsmen, stopped the flooding with only a foot of freeboard (vertical distance from the deck of the boat to the water) remaining. *Maury* arrived on scene, and dewatering was completed with the boat supported by the ship's large after boom. There were no personnel casualties, and Lt. James V. Dunn (*Maury*'s hydrographic officer who was also aboard the soundboat) credited prompt action by the crew with the assistance of *Point Kennedy*, in saving the boat.[9]

Time spent at Vung Tau was not all work; *Maury*'s officers and crew were also able to avail themselves of some leisure time ashore.

Photo 7-5

Beachside bars in Vung Tau, South Vietnam.
USS *Maury* (AGS-16) Kwajalein-Vietnam Survey 1966-1967 cruise book

Photo 7-6

Main street in Vung Tau.
USS *Maury* (AGS-16) Kwajalein-Vietnam Survey 1966-1967 cruise book

The Mekong Delta survey was completed on time, allowing *Maury* to proceed to Subic Bay for some additional R&R (rest and relaxation).

NHA TRANG HARBOR

Upon concluding R&R in Subic, *Maury* returned to the combat zone, this time to Nha Trang, which (as shown on the map on the following page) lay north of Cam Ranh Bay. Long hours round the clock by her soundboats and, during the last few days, by the ship, enabled the short but fruitful harbor survey to be completed in record time. While at Nha Trang, all hands took advantage of the beach parties and liberty available in the relatively secure area.[10]

Photo 7-7

Views of Nha Trang and of some *Maury* crewmembers enjoying a beach party ashore.
USS *Maury* (AGS-16) Kwajalein-Vietnam Survey 1966-1967 cruise book

Maury left the Nha Trang area on 19 July for Subic Bay.[11]

Map 7-1

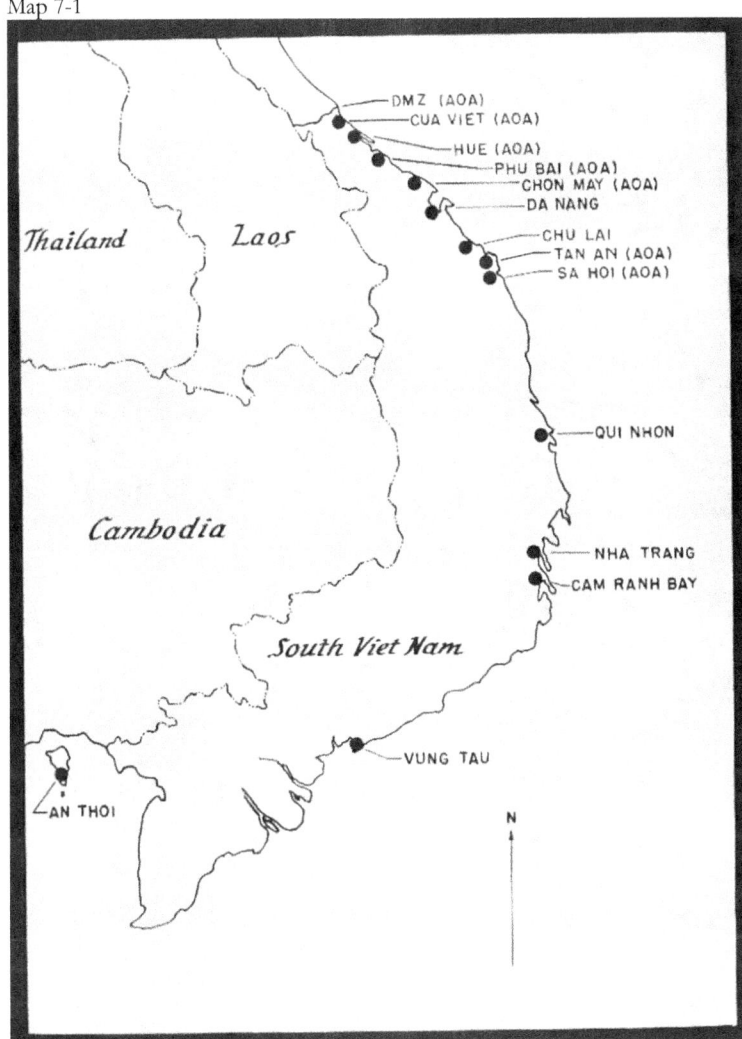

South Vietnam, and surrounding areas
USS *Eldorado* (AGC-11) Western Pacific 1967 cruise book

Other survey work in Vietnam followed, and when late August rolled around, with all ship and boat work completed, *Maury* left the combat zone for the last time—bound for Yokosuka, Japan, for liberty and an opportunity to spend some hard-earned pay.[12]

Photo 7-8

Yokosuka nightlife.
USS *Du Pont* (DD-941) cruise book

Maury and her crew bade Sayonara to Japan on 11 September, and set an easterly course, bound for Pearl Harbor.[13]

8

Final Survey Work in Vietnam

Photo 8-1

Survey ship *Tanner*, with *Sheldrake* (AGS-19) and a sound boat alongside her.
USS *Tanner* (AGS-15) 1967-68 cruise book

The remaining three of the six survey ships that served in Vietnam—*Tanner* (AGS-15), *Towhee* (AGS-28), and *Sheldrake* (AGS-19)—deployed together as a task unit in 1967-1968.[1]

Tanner left Pearl Harbor in company with *Towhee* on 18 September 1967, and returned to home port on 24 April 1968. As shown by the western portion of her track chart, she stopped at Guam and Subic en route to Vietnam; and also enjoyed crew liberty in Hong Kong and Yokosuka on the return passage to Hawaii. During the deployment, she alternated survey work in the Gulf of Thailand, with periods at Subic for upkeep and crew rest.[2]

Arriving off the western coast of Vietnam for survey operations during the latter part of October, *Towhee*'s operations continued into December. Her crew enjoyed Bangkok, Thailand, over Christmas 1967. In April, *Towhee* set a homeward course for Hawaii. *Tanner*, *Towhee*, and *Sheldrake* were each awarded a Meritorious Unit Commendation for the period 19 October 1967 to 4 April 1968.[3]

Map 8-1

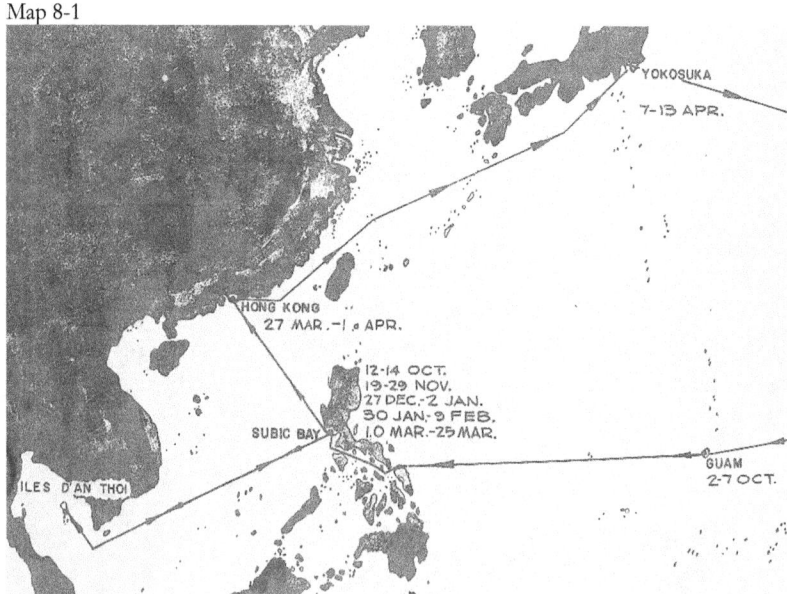

Western Pacific
Track chart from USS *Tanner* (AGS 15) 1967-1968 cruise book

During the deployment, Rear Adm. Norvell G. Ward, USN visited *Tanner*. He had recently assumed duties as commander, Service Group Three, after serving as commander, Naval Forces Vietnam. Ward was the administrative commander of Group 3, and also "double-hatted" as commander Task Force 73 (the Logistic Support Force). This task force provided the fleet with ammunition, petroleum products, and supplies, as well as many services. These included communications, towing, salvage, port service, postal, and medical support, and the movie reels passed from ship to ship.[4]

Photo 8-2

Rear Adm. Norvell G. Ward, USN, commander, Service Group Three, studies a chart during an official visit aboard the survey ship *Tanner*.
USS *Tanner* (AGS-15) 1967-68 cruise book

Entertainment by Mamie Van Doren was likely one of the high points of the cruise for many sailors. The American actress, model, singer, and sex symbol made two trips on her own to Vietnam, independent of the USO, to perform for the troops. She toured for three months in 1968, and again in 1970.[5]

Photo 8-3

Ensign Bruce, *Tanner*'s disbursing officer, enjoys attention from Mamie Van Doren. USS *Tanner* (AGS-15) 1967-68 cruise book

***MAURY* AND *SERRANO*'S FINAL OPS IN VIETNAM**

Maury departed Pearl Harbor on 29 March 1968 for her final tour in South Vietnam, in which she surveyed the area south of the Mekong Delta, while *Serrano* worked the area north of Vung Tau. *Maury* returned to Pearl Harbor on 11 October 1968.[6]

9

Seaplane Surveillance Operations

> *We covered tracks off the coast from north of Hainan Island all the way down to Da Nang. We had a bird in the air over these areas 24 hours a day, seven days a week. We saw everything from trawlers [electronics surveillance ships] to junks to steel-hulled freighters trying to get through.*
>
> —Bruce Barth, an aircrewman with VP-40 (Patrol Squadron 40) from 1965 to 1967, describing surveillance by SP-5B Marlin seaplanes of shipping routes in the South China Sea, as part of Operation MARKET TIME. Upon detecting a suspicious contact, an aircraft would descend to rig (visually inspect) the ship and, if necessary, call in air or surface assets to stop an enemy vessel. Barth also recalled "taking photos of guys ripping the covers off their ship's deck guns." Fortunately, although a few Marlins were hit by hostile fire, none were lost to enemy action.[1]

Photo 9-1

An SP-5B Marlin seaplane uses JATO (jet assisted take-off) rockets to lift from the waters of Cam Ranh Bay, South Vietnam, to begin a patrol of the coast, in April 1967. The plane was attached to VP-40, the last seaplane squadron deployed in Southeast Asia. National Archives photograph #USN 1122402

64   Chapter 9

During the Vietnam War, the Navy's Patrol Force, Seventh Fleet, maintained a continuous vigil over the waters bordering Communist countries in the Western Pacific. The headquarters for this far-flung command was, at any given time, aboard either USS *Currituck* (AV-7), *Pine Island* (AV-12), or *Salisbury Sound* (AV-13). The Pacific Fleet's three seaplane tenders alternated serving as flagship for commander, Patrol Force, Seventh Fleet—who also had the titles: commander, U.S. Taiwan Patrol Force, and commander, Fleet Air Wing One.[2]

Photo 9-2

Seaplane tender USS *Pine Island* hoisting a seaplane aboard for maintenance.
USS *Pine Island* (AV-12) Far East 1965-1966 cruise book

This arrangement charged one commander with several important and related responsibilities. These were: the anti-submarine protection of the Seventh Fleet and friendly nations in the Far East; the security of the sensitive Taiwan Strait; and overseeing the operational readiness, training, and logistics of patrol squadrons from Japan to Vietnam. The three tenders were decommissioned in 1967, following war service from

1965 into 1967. They were replaced (along with seaplanes) by newer, larger, and more capable land-based patrol planes.³

Photo 9-3

Rear Adm. Roy Maurice Isaman, USN, commander Patrol Force, U.S. Seventh Fleet. USS *Pine Island* (AV-12) Far East 1965-1966 cruise book

VIETNAM SERVICE OF THE SEAPLANE TENDERS

Seaplane tender duty in South Vietnam began before formal entry of the United States into the war. These periods of service were recognized by the awarding of Armed Forces Expeditionary Medals (AE in the following table). The association of (VS) with subsequent entries identifies periods of service at Cam Ranh Bay, for which the ships were eligible for Vietnam Service Medals, and subsequent awards of it.

Seaplane Tender Vietnam Service

Ship	Comm/ Decom	Vietnam Service Dates	Location
Currituck (AV-7)	26 Jun 44/ 31 Oct 67	6-14 Jun 64	Saigon visit (AE)
		15 May-3 Jul 65	Con Son, Con Dao Islands (AE)
		4 Jul-4 Aug 65	Cam Ranh Bay (VS)
		2-26 Sep 66	Cam Ranh Bay (VS)
		30 Nov-18 Dec 66	Cam Ranh Bay (VS)
		17 Jan-6 Feb 67	Cam Ranh Bay (VS)
		19 Mar-13 Apr 67	Cam Ranh Bay (VS)
Pine Island (AV-12)	26 Apr 45/ 16 Jun 67	4 Aug-3 Sep 64	Da Nang (AE)
		10 Oct-4 Nov 65	Cam Ranh Bay (VS)
		28 Nov-12 Dec 65	Cam Ranh Bay (VS)
		15 Jan-14 Feb 66	Cam Ranh Bay (VS)
		2-11 Apr 66	Cam Ranh Bay (VS)
Salisbury Sound (AV-13)	26 Nov 45/ 31 Mar 67	12-19 Feb 65	Da Nang (AE)
		11 May-5 Jun 65	Cu Lao Cham Island (AE)
		4-26 Mar 66	Cam Ranh Bay (VS)
		15 May-3 Jun 66	Cam Ranh Bay (VS)
		10 Jul-9 Aug 66	Cam Ranh Bay (VS)
		7-28 Oct 66	Cam Ranh Bay (VS)

Currituck made an official visit to Saigon in June 1964. That August, *Pine Island* established a seadrome at Da Nang. Because a seadrome is a landing area on the water, from which aircraft may land and take off, this was easily accomplished by her presence at the coastal city in central Vietnam, Early the following year, *Salisbury Sound* established a seadrome at Da Nang in February 1965, followed by one at Cu Lao Cham (Cham) Island, off Hoi An south of Da Nang.[4]

While *Salisbury Sound* oversaw a seadrome at Cham Island from 11 May to 5 June 1965, from which her aircraft flew patrols over the South China Sea, *Currituck* was serving the same purpose at Con Son (the largest island of the Con Dao Archipelago, and a Vietnamese penal colony), off southern Vietnam.[5]

Photo 9-4

Seaplane tender USS *Currituck* (AV-7) off Con Son in 1965. The arrival of the tank landing ship USS *Floyd County* (LST-762), accompanied by nine U.S. Coast Guard patrol boats, meant the end of her 62-day deployment to the Vietnamese island, where she had been tending patrol aircraft, and also functioning as flagship for the commander of MARKET TIME operations.
Naval Aviation News, October 1965

TENDER OPERATIONS AT CAM RANH BAY

Photo 9-5

SP-5B Marlin seaplane resting on the water in Cam Ranh Bay, South Vietnam.
USS *Pine Island* (AV-12) Far East 1965-1966 cruise book

Between 4 July 1965 and 13 April 1967, the seaplane tenders alternately served at Cam Ranh Bay, operating around the clock in support of Martin SP-5B Marlin seaplanes. While under the operational control of their parent tender, these aircraft conducted shipping, anti-submarine, and junk surveillance in South Vietnamese waters. Cam Ranh-based tenders also provided fuel, arms, and engine repairs for reconnaissance and anti-submarine seaplanes patrolling the Taiwan Straits.[6]

Photo 9-6

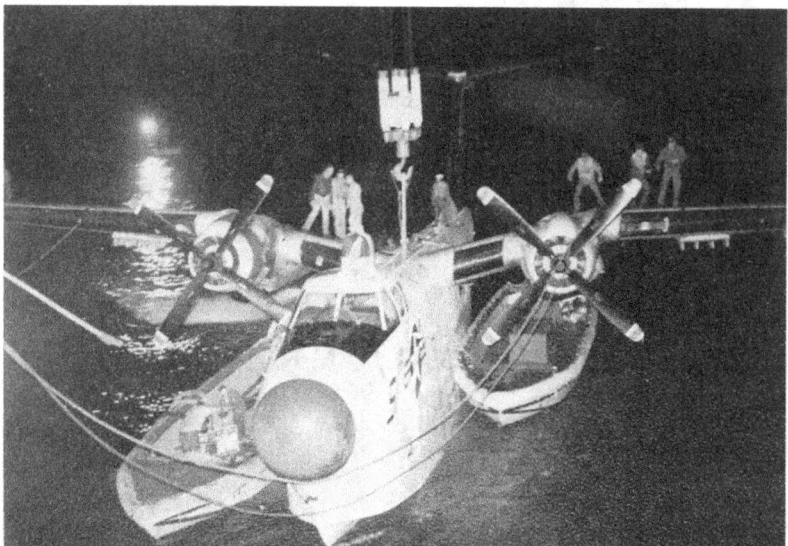

A Martin SP-5B Marlin in the water alongside the seaplane tender USS *Pine Island*.
USS *Pine Island* AV-12 Far East 1965-1966 cruise book

TENDER AND AIRCRAFT CHARACTERISTICS

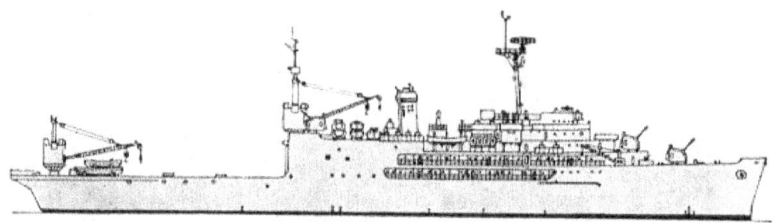

Salisbury Sound, representative of the other 540-foot *Currituck*-class seaplane tenders, displaced 15,092 tons, was broad of beam (69 feet), and deep of draft (22 feet). Propelled by four Babcock and Wilcox Express 400 psi boilers, providing steam to two Allis Chalmers turbines, coupled to two propellers via Falk main reduction gears, she could make 19 knots. Ship's complement was 162 officers and 1,085 men.[7]

Salisbury Sound was capable of supporting up to six Marlins at a time, providing aircraft upkeep and repair and personnel subsistence. Her services/facilities included engine, hydraulic, and carburetor repair, and metal, parachute, and photographic shops. In addition to her own officers and crew, she was able to billet an embarked squadron's officers and men. Her large after-deck allowed the servicing of two seaplanes at the same time, hoisted aboard by enormous cranes, one on her after-deck and one on her superstructure. Boats carried aboard refueled planes at sea and, if necessary, towed them to safety.[8]

The distinct appearance of the Marlin made it easily recognizable. Its designers had placed the engines and propellers high above spray created by water takeoffs and landings. The Marlin's bulky fuselage was perched on auxiliary wheels (used only for beaching), and the piston-type engines, slung under the gull-shaped wings, seemed too small to pull the bulky aircraft into the air. However, the seaplane's ungainly appearance belied its abilities. Cruising at 130 knots, the Marlin could employ powerful radar in its nose, and MAD (magnetic anomaly detection) gear protruding from its tail to great advantage. Its slow speed also permitted accurate drops of sonobuoys (another means of detecting submarines), and laying weapons right on target.[9]

Each plane could carry 8,000 pounds of bombs, torpedoes, and depth charges in weapons bays, located in the nacelles (streamlined housings) behind the port and starboard engines. However, while crews honed their anti-submarine warfare skills in training exercises involving U.S. submarines, rarely did a crew on patrol pick up a Soviet submarine. Marlin squadrons spent most of their time on sea surveillance, and the planes in Vietnam were fitted with rocket pods under their wings, and machine guns on each side of the fuselage.[10]

SURVEILLANCE/INTERDICTION OPERATIONS

Machine guns proved of great value against vessels running arms, munitions, or supplies to the Viet Cong in the south. Raymond T. West, who served with Patrol Squadron 50 in 1966-1967, described their usage during surveillance flights off Vietnam:

> We were deployed to Sangley Point, Philippines with...TAD (Temporary Additional Duty) detachment to Cam Ranh Bay.... There we operated off a sea plane tender; in our case the USS *Currituck* (AV-7). We were involved in Operation Market Time, the interdiction of supplies coming down the coast of North Vietnam

into South Vietnam. The SP-5B Marlin would take off and fly a four hour round-trip patrol north of Cam Ranh Bay or a four hour round-trip patrol south of the base.

The SP-2Hs, SP-5Bs, and P-3Bs [Lockheed Neptune, Marlin, and Orion patrol aircraft] armed with 2.75 folded fin rockets were hard pressed to knock off the small sampan, loaded to the gunnels with supplies and equipped with a large outboard. They were very fast and nimble targets. But the SP-5Bs armed with four M60 machine guns, one at the forward and aft hatch on each side of the fuselage made short work of the spirited craft. More than one Charlie was seen abandoning ship when faced with a Marlin skimming across the wave tops heading his way. The squadron did not lose any planes or crew members in carrying out these patrols.[11]

The term "Charlie" was a reference to the enemy. "Viet Cong" was derived from the words "Viet Nam Cong-san," which means Vietnamese Communist. This, in turn, was shortened to "VC," which in the NATO phonetic alphabet is pronounced "Victor-Charlie," and gave rise to the abbreviated "Charlie" designation.

Marlin crews had begun flying missions off the coast of South Vietnam in 1964, to stop the seaborne flow of arms and ammunition to Viet Cong forces ashore. This entailed a two-hour passage from Naval Station, Sangley Point, in the Philippines, to their assigned station, running an eight-hour patrol off Vietnam, then returning to base. To shorten transit and response times, the Navy deployed seaplane tenders to Cam Ranh Bay, to maintain, fuel, and arm the Marlins, and provide messing and berthing for the plane crews.[12]

The work day did not end for some upon return to Cam Ranh Bay, and mooring to a buoy. Bruce Barth explained:

> We'd pull buoy watch for our plane while moored—three or four enlisted men and one flight officer aboard for a six- to eight-hour shift. One would keep watch from the wing, looking out for enemy swimmers in the water. Other crewmen would sack out in lashed-down sleeping bags atop the wing. We caught some gorgeous sunsets up there. But some nights, a Navy launch crew would cruise the anchorage, tossing concussion grenades in the water at random to deter swimmers. We didn't get much sleep on those watches.[13]

BUCKNER BAY, OKINAWA

During interludes from duty at Cam Ranh Bay, seaplane tenders visited Far East ports common to other Seventh Fleet ships for crew liberty, and spent time at Buckner Bay, Okinawa—their home port overseas.

Formally Nakagusuku Bay, American servicemen had begun calling it Buckner Bay in World War II, after Lt. Gen. Simon B. Buckner Jr. was killed by enemy artillery fire on 1 April 1945, during the closing days of the Battle of Okinawa, while leading the Tenth Army in an amphibious assault on the Japanese island.[14]

Photo 9-7

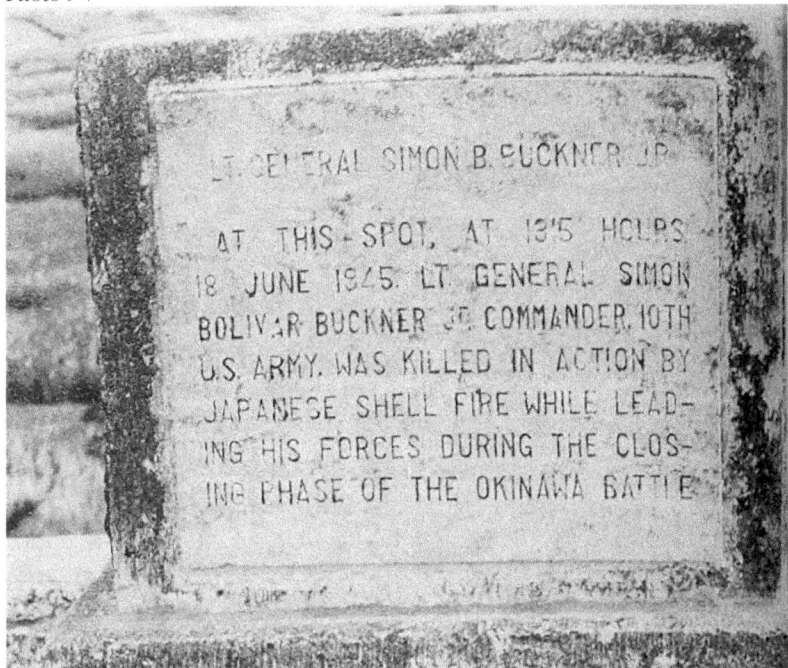

Marker identifying where Lt. Gen. Simon B. Buckner Jr., USA, was killed on Okinawa. USS *Pine Island* (AV-12) Far East 1965-1966 cruise book

END OF SEAPLANE TENDER SERVICE

By mid-1967, Lockheed P-3 Orion land-based patrol aircraft had supplanted the flying boat squadrons off the coast of Vietnam. Orions were faster, had longer ranges, could carry heavier payloads, and were cheaper to operate. Moreover, in addition to having lower operations costs, the land-based aircraft didn't face the corrosion and maintenance issues involved with sea-basing.[15]

Photo 9-8

A P-3 Orion from Patrol Squadron 47 flying a routine patrol off Cam Rahn Bay, South Vietnam, on 4 September 1968, armed with AGM-12 Bullpup missiles under its wings. National Archives photograph #K-57405

USS *Currituck* was the last Pacific Fleet seaplane tender to service SP-5B Marlins at Cam Ranh Bay, while tending aircraft of VP-40 during ongoing MARKET TIME operations. Her return to San Diego on 23 May 1967, and subsequent decommissioning on 31 October 1967, ended one of the most colorful facets of Naval Aviation operations. She was also the last seaplane tender to leave active service.[16]

10

AGTRs *Oxford* and *Jamestown*

> *I remember riding a typhoon in the South China Sea in that round bottom bucket that terrifies me to this day. Huge seas, several injuries, incredible seamanship by the "non spooks" that ran the ship.*
>
> —Former USS *Jamestown* (AGTR-3) crewmember, Cryptologist First (CTM1) Chuck Smith, recalling his tour aboard her in 1968-1969. *Jamestown* had formerly been designated AG-166.[1]

Photo 10-1

Technical research ship USS *Oxford* (AGTR-1) under way, 3 August 1964. National Archives & Records Administration photograph #KN 11023

U.S. NAVY COLD WAR INTELLIGENCE GATHERING

The Movement Report Office at the Command was a Top Secret facility [a part of the U.S. Naval Forces Philippines command, based at Sangley Point] that monitored all shipping, as well as surveillance of the movement of the Chinese government as they began occupying a majority of the atolls in the South China Sea as early as 1956.

On one particular flight a crew member of a large Chinese sampan was photographed on an atoll near Spratly Island among the many nearby reefs or atolls. The crew was filmed unloading material on the island. This appeared to be the initial incursion into the area in early 1957 since it was the first time they were seen. They also pretended to be stranded and waved accordingly. They were, of course, not present the following day and the sampan was nowhere in sight.

A majority of all "ship rigging" (photographing shipping) in the South China Sea was controlled by the naval forces command at Sangley Point, Philippines. It was not difficult determining what the ships were carrying based on a low flying pass alongside the merchant ships and photographing their laden decks. These images were later printed and sent to the United States.

—Roland Nino Martinez, PH1, USN (Retired), describing the role of Patrol Squadron VP-46 to which he was assigned in 1957 and 1958. During deployments of the squadron to Naval Air Station, Sangley Point, from San Diego, he spent much time in flight aboard P5M Martin Marlins, fulfilling his duties as squadron photographer.[2]

Photo 10-2

Naval Air Station, Sangley Point, Philippines
Courtesy of Roland Nino Martinez

Patrol aircraft surveillance of shipping off Vietnam was a continuation of well-established U.S. Navy photo-intelligence-gathering efforts run out of the Movement Report Office at Naval Air Station, Sangley Point, in the Philippines. The station was located on the northern portion of the Cavite City peninsula, approximately eight miles southwest of Manila. Although the primary mission of the squadrons operating from Sangley Point was anti-submarine warfare, aerial surveillance of shipping in the South China Sea was at the forefront in the Far East.[3]

In 1965, as the Navy began deploying seaplane tenders, and the patrol squadrons they supported, to Cam Ranh Bay, "top brass" concurrently also began sending the AGTRs *Oxford* and *Jamestown* from Subic Bay to the Gulf of Thailand to collect other forms of intelligence, by means not involving aerial surveillance. The seaplane tenders and technical research (intelligence-gathering) ships did share one thing in common. They were all of World War II vintage, originally built for duties unrelated to their current occupations.

U.S. NAVY TECHNICAL RESEARCH SHIPS

Oxford (AGTR-1) and *Jamestown* (AGTR-3) were 441-foot, converted Liberty ships, whose purpose was to "conduct research in the reception of electromagnetic propagations." In actuality, they and sister ship *Georgetown* (AGTR-2) were Cold War "spy ships." The Navy employed technical research ships during the 1960s to gather intelligence by means of monitoring, recording, and analyzing electronic communications in various parts of the world. Their mission was covert and any discussion of it was prohibited. Officially, the ships were used to conduct research into atmospheric and communications phenomena.[4]

Having been constructed as cargo ships (completed on 14 August, and 31 August 1945, respectively), *Jamestown* and *Oxford* were slow and old. Their self-defense capability was equally unimpressive; four .50-caliber machine guns comprised their only armament. Russ Mann, who served aboard *Jamestown* in 1965-1966, later remarked on her paucity of speed and armament:

> My first cruise was to South America. At that time, we had two .50 caliber machine guns. As we were returning back through the Panama Canal towards Norfolk [Virginia], we were told we were going in to be refitted with more armament and going to Vietnam. Our additional armament was two more .50 caliber machine guns! Not much defense for a ship that we joked was the only ship in the Navy that could tie up [berth alongside a pier] going full speed.[5]

The crew complement of each ship was more robust than its other characteristics, as shown in the table. Aboard each intelligence-gathering ship was a normal crew of adequate numbers to operate and maintain the vessel, as well as a full team of cryptologists to carry out its mission.[6]

USS *Oxford* (AGTR-1) (ex-SS *Samuel R. Ailken*)		USS *Jamestown* (AGTR-3) (ex-SS *J. Howland Gardner*)	
Length: 441 feet	Speed: 11 knots	Length: 441 feet	Speed: 11 knots
Displ: 11,365 tons	Complement: 254	Displ: 11,375 tons	Complement: 313
Beam: 59 feet.	Armament: four .50-caliber MGs	Beam: 56 feet	Armament: four .50-caliber MGs
Draft: 22 feet	Propulsion: steam	Draft: 27 feet	Propulsion: steam

VIETNAM DUTY

Home ported at Norfolk, Virginia, for duty with the Service Force, Atlantic Fleet, *Oxford* and *Jamestown* were transferred to the Pacific Fleet, following a decision to do so by the United States Intelligence Board in April 1965. President Dwight D. Eisenhower had set up the board in 1957, to provide a consolidated means for intelligence chiefs across the various intelligence bodies in the U.S. government, to provide advice to the director of Central Intelligence.[7]

USS *Oxford* received a message on 26 May 1965, transferring her to the Pacific Fleet. She arrived in the Western Pacific and on 16 June, stood out of Subic Bay, bound for the South China Sea. This pattern of operations—multiple, short operating periods in the combat zone—would stretch into autumn 1969 for her and *Jamestown*. Collectively, the two ships racked up fifty-three qualifying periods for receipt of the Vietnam Service Medal.[8]

- *Oxford*: 31 periods between 4 July 1965 and 23 October 1969
- *Jamestown*: 22 periods between 7 January 1966 and 23 October 1969

Map 10-1

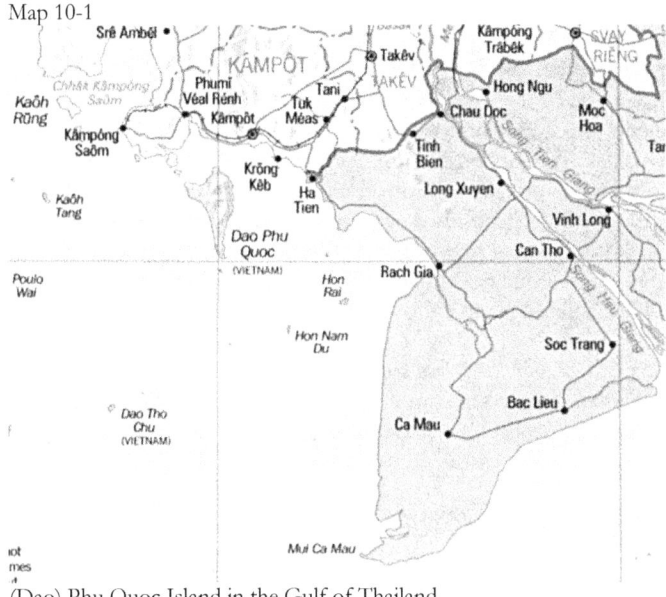

(Dao) Phu Quoc Island in the Gulf of Thailand

Oxford and *Jamestown* supplemented land-based radio-intelligence monitoring of North and South Vietnamese and Cambodian VHF and HF communication nets, until the ships were decommissioned on or about 19 December 1969 at Yokosuka. The following description of their activities is from *Cryptologic History Series, Southeast Asia, Focus on Cambodia*, January 1974, at Part One, pages 59-60. (When the formerly Top Secret, NSA book was declassified, some material was redacted.)

> In 1965, two Technical Reconnaissance Ships, the USS *Oxford* and USS *Jamestown*, began to undertake SIGINT [Signals Intelligence] tasks in the waters off South Vietnam. Formerly Liberty ships of World War II vintage and re-designated as TRSs in the early 1960's, the ships were to serve as a contingency force in the event that land-based SIGINT producing sites in Southeast Asia should be lost to the enemy and to undertake developmental collection missions. Following a United States Intelligence Board decision in April 1965, the two ships moved, accordingly, from their operational areas [redacted] coasts to the Western Pacific. The USS *Oxford* deployed to Southeast Asia waters in May 1965, and the USS *Jamestown* arrived in the Western Pacific in December 1965. From these dates until their deactivation and decommissioning in mid-December 1969, the *Oxford* and *Jamestown* engaged in a wide variety of collection tasks in Southeast Asia. Primary collection targets, among others, were

North Vietnamese, South Vietnamese, and Cambodian communications.

In the execution of their assigned missions, the research operations departments on *Oxford* and *Jamestown* undertook collection tasks, processed or recorded the raw intercept for further analysis and reported their product through the cryptologic chain-of-command. Shipboard personnel performed cryptanalysis, traffic analysis, signal analysis, translation, and transcription. Modes of intercept included manual Morse, automatic Morse, radiotelephone, single sideband and double single sideband radiotelephone, [redacted].

The TRSs *Oxford* and *Jamestown* rotated off Phu Quoc Island (near the Cambodian/South Vietnam coast) to collect SIGINT search, development, and collection of Cambodian communications. The USS *Jamestown* was successful in intercepting Cambodian [redacted].

IMPORTANT EFFORTS RECOGNIZED

The technical research ships *Oxford* and *Jamestown* received a Meritorious Unit Commendation for their considerable contributions to the war effort between 1965 and 1969. The associated citation, signed by Adm. Elmo R. Zumwalt Jr., USN, chief of Naval Operations, reads in part:

> For meritorious service from 1 November 1965 to 30 June 1969 while participating in combat support operations in Southeast Asia. Through research and the compilation of extremely valuable technical data, USS *Jamestown* and USS *Oxford* contributed most significantly to the overall security of the United States and other Free World forces operating in support of the Republic of Vietnam.

11

Major Communications Relay Ships

> *She is a floating communications station, able to stay beyond the reach of hostile powers, not depending upon the whims of reluctant allies, carrying vital communications wherever a ship can go, and capable of reaching any land area with her powerful transmitters. As her motto states, she is truly "VOX MARIS," the "Voice of the Sea" and the Voice from the Sea: the Voice of Naval Command, the Sound of United States Seapower.*
>
> —USS *Annapolis* (AGMR-1) Western Pacific 1967 cruise book

Photo 11-1

USS *Annapolis* (AGMR-1) under way in New York Harbor, 12 June 1964, soon after her conversion from escort carrier USS *Gilbert Islands* (CVE-107). Naval History and Heritage Command photograph #NH 106715

In August of 1962, the decommissioned escort aircraft carrier USS *Gilbert Islands* (CVE-107) was towed from her berth at Bayonne, New Jersey, to the New York Naval Shipyard for conversion to AGMR-1— the Navy's first major communications relay ship. She had seen action in World War II at Okinawa and Borneo. Now the nineteen-year-old ship was to be converted to fill a vital communications requirement of the Navy's operational forces.¹

The conversion from CVE to AGMR involved the modification of her flight deck to include a hurricane bow, and replacement of WWII armament with four twin 3-inch/50 caliber anti-aircraft gun mounts, two per side. The flight deck was converted to an antenna array via the installation of two directional and two omni-directional antennas. The aircraft hangar bay was converted into communication spaces to house twenty-four radio transmitters with low through ultra-high frequencies. One aircraft elevator was retained to allow servicing of equipment and boat storage.²

In addition to her other considerable communications capabilities, *Annapolis* was the first Navy fleet unit with satellite communications, following delivery of the new technology (equipment and satellite dish antenna) to her in late 1966, while at U.S. Naval Base, Subic Bay. USNS *Kingsport* (T-AG-164) had earlier been fitted with a 30-foot parabolic antenna encased in a 53-foot radome, but her rudimentary equipment proved unsuitable for combat fleet operations.³

DUTY IN VIETNAM

The Navy put *Annapolis* into service to receive, process, and deliver vital communications between the Pentagon and commanders in the Vietnam theater of operations. Not surprisingly, nearly all of her active service was spent off Vietnam, while logging twenty tours of duty in the war zone between 15 September 1965 and 29 March 1969. During interludes "off the line," *Annapolis* called at Yokosuka, Japan for dry-docking and maintenance, and other ports for crew liberty. These included Sasebo, Japan; Hong Kong; Kaohsiung, Taiwan; Singapore; Perth, Australia; Keelung, Taiwan; and Uruma (White Beach), Okinawa. However, the majority of her time in port was spent at Subic Bay.⁴

While on station off the coast of Vietnam, *Annapolis* commonly anchored for a few hours, every ten to fifteen days, outside Cam Ranh Bay, to receive mail and transfer priority crew. During these brief stops, Navy PCFs (fast patrol boats, termed "Swift boats") would often come alongside to receive much appreciated ice cream in 3-gallon containers.⁵

Arriving off the coast of Vietnam in September 1965, *Annapolis* immediately began providing communication services between naval units and shore communication facilities. In late 1966, the first ship-to-shore satellite radio message ever transmitted and received was between *Annapolis* in the South China Sea and the Pacific Fleet headquarters at Pearl Harbor. Until *Arlington* (AGMR-2) joined her in Vietnam in the latter part of 1967, *Annapolis* averaged fifty-five days at sea between port calls due to high communication demands.⁶

Routine work included providing fleet broadcasts, inter-area relay circuits, and relay circuits for ship-to-ship to shore communications. Of much importance, *Annapolis* relayed Sitreps (situation reports) of the North Vietnamese military operations to the commanders in Washington DC, via satellite, with some political interest situations reported directly to the White House. Information about the Tet Offensive and the Battle of Khe Sanh, for example, were often reported in minutes via Sitreps or Progreps (progress reports), whereas relay of critical info by regular communications often took hours, at best.⁷

USS *ARLINGTON* JOINS THE FLEET

Photo 11-2

USS *Arlington* (AGMR-2) under way, circa 1967.
Naval History and Heritage Command photograph #NH 97625

USS *Arlington* was the former light fleet carrier *Saipan* (CVL-48), built on a heavy cruiser hull. Placed in commission on 14 July 1946, too late for service in World War II, *Saipan* was decommissioned in 1957. Her conversion to major communications relay ship began on 1 September 1964 and was completed in July 1966. *Arlington* was commissioned on 27 August 1966. Like *Annapolis*, her armament consisted of four 3-

inch/50 twin mounts. Ship's manning was also similar, with 45 officers and 882 enlisted men comprising her crew complement.[8]

Arlington's tours of duty off Vietnam began on 22 August 1967, and ended 8 July 1969—twelve in total over a nearly two-year period. Much of her time was spent performing communications relay duties at Yankee Station in the Tonkin Gulf and, on at least one occasion, providing communications support farther south for ships in the MARKET TIME area. She also supported the Apollo 8, Apollo 10, and Apollo 11 missions, as part of Task Force 130 (Manned Spacecraft Recovery Force, Pacific), in late December 1968, late May 1969, and late July 1969, respectively.

MANNED SPACECRAFT RECOVERY FORCE PACIFIC

> *USS* ARLINGTON *was used during Apollo 11 to provide full support for Presidential communications, and to provide backup communications on recovery day. For Presidential support* ARLINGTON *established two 6 kHz HF radio trunks with* NAVCOMMSTA HONO *containing three voice and six teletype circuits. Due to the deterioration of TACSAT and the poor performance of ATS, a third HF trunk was established for UHF/HF relay of CTF 130 C/C and NASA PRS COORD voice circuits.*
>
> —Description of the communications support provided by *Arlington* for President Richard M. Nixon and NASA during recovery of the Apollo 11 command module. Nixon was on scene for the return to Earth of the astronauts involved in the first moon landing.[9]

USS *Arlington*'s first involvement with the Apollo program came on 18 December 1968, when she departed Hawaii with units of Task Force 130. Acting as primary landing area communications relay ship, she participated in the recovery of Apollo 8 and returned to Pearl Harbor twelve days later. Apollo 8 (the second manned spaceflight mission) was the first astronaut-controlled craft to leave low Earth orbit, circle the moon, and return safely to terra firma. Launched on 21 December 1968 with Frank Borman, James Lovell, and William Anders on board, the capsule dropped into the Pacific on 27 December.[10]

The U.S. Navy ships assigned to the Manned Spacecraft Recovery Force, Pacific, for the Apollo 8, 10, and 11 missions are identified in the table. The letters "PRS" denote Primary Recovery Ship.

| Apollo 8 | Apollo 10 | Apollo 11 |
21-27 December 1968	18-26 May 1969	16-24 July 1969
Yorktown (CVS-10) PRS	*Princeton* (LPH-5) PRS	*Hornet* (CV-12) PRS
Arlington (AGMR-2)	*Arlington* (AGMR-2)	*Arlington* (AGMR-2)
Chipola (AO-63)	*Carpenter* (DD-825)	*Carpenter* (DD-825)
Chuckawan (AO-100)	*Chilton* (LPA-38)	*Goldsborough* (DDG-20)
Cochrane (DDG-21)	*Ozark* (MCS-2)	*Hassayampa* (AO-145)
Francis Marion (APA-249)	*Rich* (DDE-820)	*New* (DD-818)
Guadalcanal (LPH-7)	*Salinan* (ATF-161)	*Ozark* (MCS-2)
Nicholas (DD-449)		*Salinan* (ATF-161)
Rankin (AKA-103)		
Rupertus (DD-851)		
Salinan (ATF-161)		
Sandoval (LPA-194)[11]		

Photo 11-3

USS *Arlington* as viewed by the Apollo 8 astronauts from the recovery ship.
USS *Arlington* (AGMR-2) 1968 cruise book

APOLLO 10 MISSION

On 2 May 1969, *Arlington* joined TF 130 in Hawaii, to serve once again as primary landing area communications relay ship. She departed Pearl Harbor on 11 May, bound for the Apollo 10 recovery area some 2,400 miles south of Hawaii. Splashdown occurred on the 26th in calm waters of the South Pacific less than three miles from USS *Princeton*, tasked with recovering the astronauts and their spacecraft. From *Princeton*'s deck,

Navy helicopters carried members of Underwater Demolition Team 11, to Apollo 10's command module.[12]

Photo 11-4

Descent of the Apollo 10 space capsule as photographed from USS *Arlington*.
USS *Arlington* (AGMR-2) Middle Pacific 1969 cruise book

Eight days after departing the earth for a manned mission to the moon on 18 May 1969, the command module of Apollo 10 reentered Earth's atmosphere at 32 times the speed of sound. Aboard were three astronauts: the mission commander, Col. Thomas P. Stafford, USAF; lunar module pilot, Comdr. Eugene A. Cernan, USN; and command module pilot, John W. Young, USN. Following liftoff at 1249 and a bumpy escape from the atmosphere, the mission settled into routine but highly complex maneuvers that saw the lunar module and the command service module undocked, then docked, undocked, docked, and undocked again in the course of the journey to and from the moon.[13]

Stafford, Cernan, and Young were the first astronauts to orbit the moon in a spacecraft fully capable of landing a person on the moon. From just 15 kilometers above the moon's surface, Stafford and Cernan were able to identify landing sites for future missions. This information, along with production of lunar maps by robotic probes and fulfillment of all other primary mission objectives, gave NASA planners the confidence they needed to launch Apollo 11 in July of the same year.[14]

At the site of the Apollo 10 recovery, *Arlington* functioned as a communications relay link between Naval Communications Station, Honolulu, 2,500 miles to the northeast, and *Princeton*, a few thousand yards away. Navy and NASA voice circuits were maintained at home telephone quality for the recovery. After a majestic descent, the command module set down a scant 3.2 miles from *Arlington*. For each manned space mission, the Navy assigned either an aircraft carrier (CVA or CVS) or a helicopter amphibious assault ship (LPH) as the primary recovery ship, because they carried helicopters required to lift the recovered astronauts and spacecraft aboard. Navy UDT swimmers were dropped into the water by helicopters to assist with the recovery operations.[15]

Her support of the operation completed, *Arlington* proceeded to Midway Island, to provide communications support for a conference on 8 June 1969 between Ricard M. Nixon and President of South Vietnam Nguyen Van Thieu.[16]

NIXON-THIEU CONFERENCE AT MIDWAY ISLAND

> *The war in Vietnam concerns not only Vietnam but the entire Pacific. The people of South Vietnam, however, have the greatest stake. If the peace is inadequate, there will be repercussions all over Asia. There can be no reward for those engaged in aggression. At the same time, self-determination is not only in the Vietnamese interest, but in the American interest as well. It would improve the prospects of peace throughout the Pacific.*
>
> —Statement by President Richard M. Nixon to South Vietnamese president Nguyen Van Thieu, at the Midway Island conference on 8 June 1969, while emphasizing the importance of ending the Vietnam War honorably.[17]

86   Chapter 11

Although Midway had adequate naval communications for ordinary message traffic, the Pacific Fleet headquarters sent *Arlington* to the isolated island (located more than 1,000 miles to the northwest of Hawaii) to provide additional circuits for the influx of correspondents covering the conference. This was accomplished by establishing, 10 two-way circuits for relay via her powerful transmitters and sensitive receivers, that were channeled at Honolulu into regular commercial circuits to the continental United States.[18]

Photo 11-5

U.S. Naval Air Station, Midway.
USS *Arlington* (AGMR-2) Middle Pacific 1969 cruise book

Photo 11-6

President Richard M. Nixon and entourage and debarking from Airforce One.
USS *Arlington* (AGMR-2) Middle Pacific 1969 cruise book

At the meeting on 8 June, Nixon and Thieu discussed the withdrawal of U.S. troops from Vietnam and U.S. negotiating strategy with the North Vietnamese at the Paris Peace Talks. Following the meeting, Nixon announced the impending scheduled withdrawal of 25,000 American troops. This action was in concert with the administration's Vietnamization policy, enacted soon after President Nixon took office in January 1969. The plan was to train, equip and expand South Vietnamese forces so that they could take over more military responsibilities for their own defense against the North Vietnamese Communists. It was believed this action would allow, at the same time, the U.S. to gradually withdraw its combat troops from South Vietnam, which were then at 475,200 personnel.[19]

Photo 11-7

President Nixon and President Thieu shaking hands.
USS *Arlington* (AGMR-2) Middle Pacific 1969 cruise book

When Nixon took office, U.S. combat troops had already been fighting in Vietnam for nearly four years since 1965, and some 31,000 had lost their lives. Nonetheless, that military support and commitment had made little progress in defeating North Vietnamese troops and the Viet Cong. Under continuing fierce and intense protests, Nixon and his advisers sought a way to disengage U.S. combat forces without appearing to abandon South Vietnam in the war against the Communists. The strategy to do so was called Vietnamization.[20]

A battalion of the U.S. 9th Infantry Division left, on 7 July 1969, in the initial withdrawal of U.S. troops. The 814 soldiers were the first of 25,000 troops withdrawn in the initial stage of the U.S. disengagement from the war. There would be fourteen more increments in the withdrawal, with the last U.S. troops not leaving until after the Paris Peace Accords were signed in January 1973.[21]

APOLLO 11

Photo 11-8

A Navy helicopter from Helicopter Anti-Submarine Squadron Four (HS-4) picking up the astronauts from the Apollo 11 command module on 24 July 1969.
Naval History and Heritage Command photograph

The following month, Apollo 11 was launched on 16 July 1969 from the Kennedy Space Center in Florida, atop a massive Saturn V rocket. Neil

Armstrong was the mission commander, Edwin "Buzz" Aldrin the lunar excursion module ("Eagle") pilot, and Michael Collins the command module ("Columbia") pilot. Armstrong was a naval officer, while Aldrin and Collins were both U.S. Air Force officers. Armstrong piloted the lunar module to the moon's surface on 20 July, with Aldrin aboard. Collins remained aboard the command module.[22]

Armstrong exited the lunar module at 1056, and exclaimed, "That's one small step for [a] man, one giant leap for mankind," as he made his famous first step on the moon. Armstrong and Aldrin then collected samples, conducted experiments, and took photographs, including ones of their own footprints. Returning to earth on 24 July, the Apollo 11 craft came down in the Pacific, west of Hawaii. The crew and the craft were picked up by the *Hornet*, and the three astronauts were put into quarantine for three weeks.[23]

Photo 11-9

Armstrong, Aldrin, and Collins enter the Mobile Quarantine Facility aboard USS *Hornet* (CVS-12), following the recovery of Apollo 11 on 24 July 1969. Naval History and Heritage Command photograph

Arlington had arrived in the recovery area (about 1,100 miles southwest of Oahu) on 21 July, tested her equipment and, the following day, moved to Johnston Island. She embarked President Nixon for an overnight visit on the 23rd. Nixon, and an entourage that included Secretary of State William P. Rogers and National Security Advisor Henry Kissinger, had flown on Air Force One to Johnston Atoll, and then on Marine One to the command ship *Arlington*. On the 24th, the party took the helicopter to *Hornet*, and were greeted by Adm. John S. McCain, Jr., the commander in chief, Pacific Command.[24]

With crew and capsule successfully recovered, and the astronauts put into the mobile quarantine facility aboard *Hornet* and welcomed back by President Nixon through its window, *Arlington* headed for Hawaii. From there, she proceeded to Long Beach, California, arriving there on 21 August. Four days later, she moved south to San Diego to begin inactivation. She was decommissioned on 14 January 1970 and joined the Inactive Fleet at San Diego.[25]

Annapolis had been decommissioned earlier on 20 December 1969 at Naval Station Norfolk; transferred to the Atlantic Reserve Fleet; and towed to the Philadelphia Naval Shipyard, where she was "placed in mothballs" (laid up).[26]

LAURELS FOR THE COMMUNICATIONS SHIPS

USS *Arlington* earned three Meritorious Unit Commendations (MUCs) for her support of the Apollo 8, Apollo 10, and Apollo 11 missions, respectively. The associated eligibility periods were 11-29 December 1968, 4-31 May 1969, and 20-25 July 1969.

USS *Annapolis* was awarded a MUC by Rear Adm. Robert L. J. Long, commander, Service Group Three, for the period 9 January 1967 to 29 January 1968. Until August 1967, *Annapolis* had operated in the waters off the coast of South Vietnam, providing support to surveillance units in the MARKET TIME area. In August, with the arrival of *Arlington* and the commissioning of Naval Communications Station, Cam Ranh Bay, *Annapolis* moved to a position near the entrance to the Tonkin Gulf. From there, she provided communications services to Amphibious Ready Groups, Carrier Strike Groups, and ships involved in gunfire support and SEA DRAGON missions against North Vietnamese coastal targets.[27]

12

Patapsco-class Gasoline Tankers

Operations in Vietnam, while never quite routine, became at least somewhat predictable on the first cruise in 1966. In Da Nang, NOXUBEE would usually go alongside a USNS tanker, frequently USNS SAUMICO, to load POL [Petroleum, oil and lubricants], we would then go anchor for 2 to 3 days before sailing. Usually our destination would be Cua Viet where we would anchor and pump through a floating hose tied to a buoy. One benefit of being the Liquid Cargo Officer was being invited aboard SAUMICO by the Chief Mate to have a beer. Please remember, there are two kinds of beer: good beer and great beer. Great beer is cold beer. This was especially so in Vietnam, Republic of. I also recall that most of the Wardroom was sometimes invited over and also the Chief's Mess. I distinctly remember carrying back a ditty bag of great beer on a few occasions and giving most of it to the Chief's Mess where it would do the most good.

—Rik Kuhn recalling duty in Vietnam as a junior officer aboard the gasoline tanker USS *Noxubee* (AOG-56).[1]

Photo 12-1

Gasoline tanker USS *Noxubee* (AOG-56) under way.
Naval History and Heritage Command photograph #L45-209.06.02

USS *Noxubee* was one of the twenty-three *Patapsco*-class gasoline tankers (AOG-1 through AOG-11, and AOG-48 to AOG-59) built for the Navy during World War II. Six of those still in service—or brought out of "mothballs" and pressed into service—served in Vietnam. As indicated by the "Naval Service" column in the table, *Elkhorn* remained in service from commissioning on 22 February 1944, until she was decommissioned on 1 July 1972. On the other hand, *Patapsco* was put out of service after World War II, recommissioned for duty in the Korean War, decommissioned a second time, and then brought back a third time for duty in Vietnam. (The acronym MARAD in the table is short for, United States Maritime Administration.)

Patapsco-class Gasoline Tankers that served in Vietnam

Ship	Naval Service	Disposition
Elkhorn (AOG-7)	12 Feb 44-1 Jul 72	Sold to Taiwan, renamed ROCS *Hsing Lung* (AOG-515)
Genesee (AOG-8)	27 May 44-16 Dec 49 28 Jul 50-5 Jul 73	Transferred to Chile under the Security Assistance Program, renamed *Beagle*
Kishwaukee (AOG-9)	27 May 44-2 Apr 58 1 Sep 66-15 Jan 70	Sold by MARAD to Mid Pacific Sea Harvesters Inc., for conversion to a fishing vessel
Noxubee (AOG-56)	19 Oct 45-6 Mar 59 10 Sep 66-1975	Sold by MARAD 24 Mar 76 for scrapping
Patapsco (AOG-1)	4 Feb 43-29 May 46 19 Oct 50-29 Jun 55 18 Jun 66-unknown	Struck 1 Aug 74; sold by MARAD 1 Nov 79; became the fishing trawler *Artic Stor*
Tombigbee (AOG-11)	13 Jul 44-12 Dec 49 28 Jul 50-7 Jul 72	Sold to Greece under the Security Assistance Program, renamed HRNS *Ariadni* (A-414)[2]

Dramatically increased military operations in Vietnam demanded additional naval forces, and gasoline tankers were sent to the war zone as part of this effort. From March 1965 until late 1971, at least one AOG was constantly deployed in Vietnamese waters. The period in which each AOG became eligible for its first Vietnam Service Medal, and the ending date of their last award, follows:

- *Elkhorn*: 12 November 1965 – 26 January 1970
- *Genesee*: 4 July 1965 – 16 October 1968
- *Kishwaukee*: 31 December 1966 – 18 May 1968
- *Noxubee*: 5 April 1967 – 9 January 1970
- *Patapsco*: 22 October 1966 – 22 June 1969
- *Tombigbee*: 1 April 1966 – 8 November 1971[3]

REPRESENTATIVE OPERATIONS

A deployment by USS *Kishwaukee* from 1 November 1967 to 15 May 1968 is representative of those by other AOGs during the war. She (like her sisters) was a unit of Service Squadron Five, based in Hawaii. Carrying hundreds of thousands of gallons of volatile gasoline, the tankers wisely did not seek to employ their 3-inch/50 guns in battles with enemy forces ashore. The Viet Cong and North Vietnamese did not share this view. Four AOGs—*Genesee*, *Kishwaukee*, *Noxubee*, and *Patapsco*—earned combat action ribbons, with *Noxubee* and *Patapsco* garnering two apiece. *Kishwaukee*'s encounter with North Vietnamese Army artillery fire is described, as well as enemy actions involving other gasoline tankers.

VIETNAM BOUND

Vietnam deployments of Navy ships based in Hawaii, began with their passage to sea. After clearing the channel at Pearl Harbor and headed fair, Diamond Head (on Waikiki's eastern coastline) faded into the distance. Smooth sailing en route to the war zone might lie ahead, but Pacific crossings could be arduous—particularly during typhoon season.

Photo 12-2

View of the channel at Pearl Harbor, Hawaii, with Diamond Head in the background. USS *Kishwaukee* (AOG-9) Western Pacific 1967-1968 cruise book

The wide flat-bottomed vessels tended to roll in any kind of sea. Full or empty, it did not seem to make much difference. Tankers even rolled when at anchor in a harbor, if not well protected from wind and sea. Their crews endured frequent soaking, there being no way to traverse the length of an AOG below decks. To get around, it was necessary to cross the open tank deck. In heavy seas, sailors often dashed across the deck to avoid crashing waves. In really dirty weather, a lifeline was rigged to provide something to hold on to. When a ship was fully loaded, the tank deck was only about three feet above the waterline. In any kind of a sea, even a slight roll of the ship allowed waves to break over the deck, roll aft and crash into the after deckhouse. The resultant spray drenched anyone trying to cross the more desirable catwalk above the deck.[4]

Following a 6,000-mile passage from Pearl, *Kishwaukee* arrived at Da Nang on 1 December 1967. Working for commander, Naval Support Activity, Da Nang, her job was to provide petroleum products to the outposts of the northern part of I Corps. Bordering the Demilitarized Zone, I Corps saw heavy fighting almost continuously from 1956 to 1975. The I Corps Tactical Zone stretched south from the DMZ through the bases along Route 9 in Quang Tri Province, down through Thua Then Province (with the Ashau Valley) and Quang Nam Province (where the first Marines landed in 1965) to the Viet Cong-infested Quang Ngai Province.[5]

Map 12-1

Abbreviated map of South Vietnam, showing only I-Corps, and part of II-Corps of the five total Corps Tactical Zones stretching north to south in the country

Each morning, while pumping to "sea lines" (fueling connections near the shore) or approaching the anchorage, *Kishwaukee* was greeted by the fishing fleets. Though the tides of war changed their lives, the Vietnamese still relied on subsistence from the sea. Fueling trips to other locations usually began with loading from a USNS tanker in Da Nang Harbor. When an emergency existed elsewhere, and no afloat source was available, *Kishwaukee* would load from storage tanks ashore.[6]

Kishwaukee taking fuel from a USNS tanker manned by a civilian crew.
USS *Kishwaukee* (AOG-9) Western Pacific 1967-1968 cruise book

Upon arrival at a shore station to deliver fuel, AOGs normally anchored 1,500 to 2,000 yards offshore, alongside a buoy marking the seaward terminus of a pipeline. A boat would then take a 4-inch hose from the ship's tank deck to the buoy, for connection to a swiveling gooseneck connection. This completed, transfer of fuel ashore could commence, a lengthy process owing to the AOGs' limited pumping capacity and small-diameter pipelines.[7]

After acquiring .50-caliber ammunition from sister ship *Tombigbee*, *Kishwaukee* stood out to sea, bound for Cua Viet up the coast, located about five miles below the DMZ. Upon arrival, she found the bottom lay system inoperative, due to the monsoon season. However, using "Mike 8" boats (landing craft) loaded with 10,000-gallon bladders, she was able to transfer 300,000 gallons of fuel ashore in less than 24 hours.[8]

Once the fuel was ashore, it was stored in 10,000-gallon neoprene bladders in sand bag enclosures. From this "tank farm" at the shoreline, the fuel was transported by various means to units requiring it. *Kishwaukee*, and other AOGs, provided fuel for Army and Air Force units, as well as servicing those of the Navy and Marine Corps.[9]

Photo 12-4

Modest storage area, partially enclosed by stacked bags of sand to protect fuel within.
USS *Kishwaukee* (AOG-9) Western Pacific 1967-1968 cruise book

Sending fuel ashore to the U.S. Army at Sa Huynh, where it had established a camp on the southern tip of I Corps, required a different method. Empty barges were towed at high tide out of the lagoon over a sand bar, and anchored. *Kishwaukee* then anchored seaward of the barges, and passed a hose for fueling. At the next high tide, the "topped off" barges were taken back into the lagoon to the camp, located on a small island.[10]

Photo 12-5

U.S. Army barge at Sa Huynh, South Vietnam.
USS *Kishwaukee* (AOG-9) Western Pacific 1967-1968 cruise book

The nearby small fishing village of Sa Huynh lay about 130 miles south of Da Nang, and below Quang Ngai on the Vietnamese coast.[11]

Photo 12-6

Fishing village of Sa Huynh, located on furthermost coastal point of I Corps, USS *Kishwaukee* (AOG-9) Western Pacific 1967-1968 cruise book

ENEMY ATTACK ON FUEL FARM AT CUA VIET

Photo 12-7

Fuel farm at Cua Viet under enemy artillery fire on 20 February 1968. USS *Kishwaukee* (AOG-9) Western Pacific 1967-1968 cruise book

The tank farm at Cua Viet was the most northern one in I Corps, closest to the DMZ, and probably the most important stop of the deployment.

Fueling operations there involved pumping through the bottom lays and occasionally offloading to a YOG (self-propelled gasoline barge). On 20 February 1968, the fuel farm came under artillery fire four times, while *Kishwaukee* was anchored about 1,500 yards away. No damage to her or crew resulted, but it reminded everyone of their vulnerability.[12]

Following fueling operations at Cua Viet, the ship returned to Da Nang, and on 23 February, she departed for Subic Bay—following ninety-three continuous days "on the line." Since the beginning of the Vietnam War, Subic Bay had offered the major support activity outside the war zone for ships of the U.S. Seventh Fleet. For *Kishwaukee*, a visit there provided opportunity to get needed repairs made, enjoy long overdue recreation, and do a little shopping.[13]

Photo 12-8

Subic Bay, on the west coast of the island of Luzon.
USS *Kishwaukee* (AOG-9) Western Pacific 1967-1968 cruise book

Photo 12-9

Spanish Gate, U.S. Naval Base, Subic Bay, Philippines.
USS *Kishwaukee* (AOG-9) Western Pacific 1967-1968 cruise book

RETURN TO THE COMBAT ZONE

Following this pleasant interlude, *Kishwaukee* returned to the line on 21 March 1968 for another fifty-one days, during which she pumped nearly six million gallons to recipients in I Corps. As before, she operated under Naval Support Activity, Da Nang—the Navy's largest overseas shore command, which sustained some 140,000 Allied fighting forces, including 80,000 Marines in I Corps. Since its establishment on 15 October 1965 to support the Third Marine Amphibious Force, it had grown from little more than an anchorage to a bustling deep-water port handling one million tons of cargo every three months. Da Nang had detachments at Dong Ha/Cua Viet, Phu Bai/Hue/Tan May, and Chu Lai. The largest was Chu Lai, with a port capacity one-sixth that of Da Nang.[14]

Photo 12-10

Da Nang Harbor on 6 January 1966. The harbor entrance control post is at the bottom of the photo; in the middle is observatory island, location of the Naval Advisory Group. Naval History and Heritage Command photograph #NH 74482

The outlying NSA detachments proved invaluable during the Tet Offensive (a coordinated series of North Vietnamese and Viet Cong attacks launched against targets all across South Vietnam on 31 January 1968), when they assumed the logistic support of isolated allied forces.

100   Chapter 12

The units at Dong Ha and Cua Viet pushed supplies and ammunition up the Cua Viet River to the 3rd Marine Division holding the line at the DMZ, while the Tan My detachment assisted troops locked in combat at Hue. The support establishment at Chu Lai supplied the 1st Marine Division while the one at Sa Huynh supplied Army troops near Duc Pho, midway between Da Nang and Qui Nhon.[15]

Upon completion of her tour of duty, *Kishwaukee* left Vietnamese waters to make the return voyage to Hawaii. Her only port call en route was a brief two days in Yokosuka, a busy time involving repairs as well as relaxation. Some of the crew were able to visit Tokyo, but most everyone wished there had been more time to shop and tour Japan.[16]

Anticipation of their impending return home filled the remaining days of the deployment. On 15 May 1968, *Kishwaukee* stood up the channel at Pearl Harbor, and proceeded to her berth.[17]

Photo 12-11

Kishwaukee berths at Pearl Harbor, amid waves of dependents on the pier and music by a Navy band welcoming her home from deployment.
USS *Kishwaukee* (AOG-9) Western Pacific 1967-1968 cruise book

GENESEE SUFFERS ATTACK WITH ONE KILLED

> *I ... was standing on the starboard side of the bridge gazing into the water when I began to see what I thought at first to be someone throwing stones some 50 yards beyond the bow of the ship. I then realized that it was incoming rounds and I beat a quick path to the Radio Shack which was about 15 feet from where I was*

standing. To make a long story short, we took about 40 rounds on our port side, in the area of officers' country in the after part of the ship and the tank deck.

—Chief Radioman John Heatherman, USN (Retired), describing an NVA artillery attack on the *Genesee* (AOG-8) on 22 April 1968. In order to carry out her mission, she lay at anchor in the mouth of the Cua Viet River, pumping fuel ashore.[18]

Photo 12-12

Naval Support Activity, Da Nang, Cua Viet detachment, October 1965.
Naval History and Heritage Command photograph #NH 74205

On 22 April 1968, while sister ship *Kishwaukee* was still serving in Vietnam, *Genesee* (AOG-8) came under attack while fulfilling an urgent need for fuel at the Marine logistics base located at the mouth of the Cua Viet River. Collocated with the Naval Support Activity, Da Nang, Cua Viet detachment, the base lay just five miles south of the DMZ. Eleven days earlier, North Vietnamese Army artillery fire had hit the base's fuel farm, destroying 40,000 gallons of petroleum.[19]

Following her arrival, *Genesee* had been preparing to anchor about a mile offshore when, due to a faulty fuel hose, the Marines requested that she move into the river mouth. Filling tanks ashore, and a fuel barge alongside her, soon commenced. Being a lengthy operation, some in the crew not involved with fueling, went ashore to watch a movie.[20]

When the gasoline tanker came under heavy enemy fire, low tide precluded safe movement of the ship. The shelling continued most of the night. During the attack, Shipfitter Fireman Arthur W. Ball was killed, and *Genesee* damaged extensively. Yeoman Chief Russell Waddell, USN (Retired) later described the circumstances under which his shipmate was mortally wounded:

> I was the last person to speak to Ball the night he died. He was coming across the catwalk and I was on fire watch [positioned with firefighting equipment, in the event of a fire while pumping fuel]. He looked up and asked me if he was ready to go. He was being sent home in two days as his time in the Navy was up. I told him he was ready to go and then we came under attack. He died dogging down the hatch he walked through after he left me.[21]

When tidal conditions allowed, *Genesee* proceeded down the coast to report her status. The HF radio antennas had been damaged, reducing communications to UHF teletype messages. However, sighting a communications van ashore, she was able to use her UHF "line of sight" radio to send a message to the shore facility for relay to the Philippines. *Genesee* proceeded under her own power to Subic Bay, where she was drydocked for forty-one days while undergoing repairs.[22]

NOXUBEE TAKEN UNDER ENEMY FIRE

On 28 October 1968, *Noxubee* (AOG-56) came under attack—like *Kishwaukee* and *Genesee* had in February and April—at Cua Viet. On that day, she was pumping to an underwater pipeline and a bladder boat alongside when, at 1530, an urgent message came into Radio Central ordering the ship to get under way immediately. The message warned that a NVA attack was imminent. Radioman Third Ed Angeloff was just beginning to copy the message, to take to the commanding officer, when artillery rounds began falling near the ship.[23]

Ens. Richard Bland, the command duty officer, was at the pumping station on the port side of the tank deck, when plumes of water began appearing off the starboard side. He quickly called away the sea detail and ordered the engine room to standby for an emergency under way. In response, the engineers started all diesel propulsion engines (and brought them up to speed), as well as additional generators to provide electrical power to the gun mounts. Gun crews did not return fire because they could not see where the enemy rounds were coming from, and there were friendly forces ashore.[24]

When the First Lieutenant Ens. Andy Bavarik and Boatswain's Mate Chief Thomas Franklin saw the splashes from the tank deck and heard the emergency under way ordered, Franklin lay aft and brought the stern anchor up off the bottom. Bavarik, and a first class boatswain's mate, went to the foc's'le and used the anchor windlass to pull *Noxubee* forward toward the bow anchor imbedded in the sea floor. Years later, Bavarik remembered the splashes being where the tanker had been.[25]

Concurrently, Bland and the tank deck crew cast off the bladder boat, and used fire axes to cut away the fueling hoses. All the while, rounds bracketed the ship, as enemy gunners searched for the range. As soon as the bow anchor was clear of the bottom, Lt. John McCall, USN, *Noxubee*'s commanding officer, had her under way at 14 knots (top speed) and safely out of range. A couple of UH-1 Huey gunships flew over the ship providing air cover. When the situation ashore stabilized, *Noxubee* returned to the Cua Viet anchorage and resumed pumping.[26]

NOXUBEE MINED OFF CUA VIET

Less than a year later, *Noxubee* (on a subsequent deployment to Vietnam with a new commanding officer, Lt. Eugene Cass, USN), was anchored off the mouth of the Cua Viet River. She had come up from Da Nang with a load of diesel fuel, jet fuel, motor fuel, and aviation fuel, and had begun pumping at 0855 on the morning of 8 September 1969.[27]

At 2137 that evening, as *Noxubee* continued to transfer fuel ashore, sentries on her fantail sighted two swimmers in the water ten to fifteen yards astern of the ship. They alerted the bridge and took the enemy under rifle fire. Seaman Paul Gryniewicz, the messenger of the watch, was on his way to the fantail when they opened fire. He got there a few seconds later, and began throwing concussion hand grenades at the two heads in the water. These type grenades (employed as deterrents against swimmer-sappers trying to affix explosives to a ship's hull) did not produce shrapnel. As such, they could be dropped overboard without worrying about possible damage to the tanker. As quickly as the swimmers had appeared, they disappeared from sight.[28]

The swimmers were "sappers," members of the Bo Doi Dac Cong ("soldiers in special forces"), a highly organized, well-trained and well-equipped organization that carried out special operations. The term "sapper" originated from the French word *saper*, a reference to French soldiers who dug narrow trenches, or "saps," toward an enemy fort to provide a somewhat protected channel for moving men and artillery closer to the fort in preparation for an assault. The term "sapper" currently refers more broadly to combat engineers who carry out a variety of construction and demolition duties. In Vietnam, American troops used the name primarily for North Vietnamese Army (NVA) and Viet Cong (VC) units that used tactics more akin to those of commandos than to the work of engineers.[29]

Prior to the Tet Offensive in early 1968, the sappers in the South were controlled by the Viet Cong and operated independently of the North Vietnamese Army. After the Viet Cong suffered massive casualties during Tet, all sapper operations in South Vietnam were supervised by the 429th Sapper Group, which reported directly to the Sapper High Command, a department in the NVA High Command in Hanoi. Training centers in South Vietnam and Cambodia were run by the 429th Sapper Group, while the centers in North Vietnam and Laos were directed from the NVA High Command. The instruction typically lasted three to eighteen months, depending on whether trainees would be soldiers in regular units or raiders operating outside a formal military structure.[30]

While anti-swimmer actions were in progress, *Noxubee*'s crew made preparations to get under way and went to their General Quarters stations expecting a mine to go off at any second. Men topside kept up a visual search for the swimmers, and small arms and .50 caliber machine gun fire, while tossing hand grenades over the side. The goal was to keep the enemy as far away as possible, if still around. The commanding officer ordered cease fire at 2203, as the ship steamed around the anchorage, waiting for a mine to go off (should there be one affixed to her hull) and divers to come out and inspect the hull.[31]

Noxubee anchored about 1,200 yards off the beach at 2215 so that EOD divers from Cua Viet could take a look. By 2350 they reported that because of darkness, strong currents and rough seas, they had been only able to inspect the stern area, and did not find anything. A more complete inspection would be made at daylight. At this point, everyone began to breathe a little easier. The swimmers had been sighted off the stern; the stern was inspected and found to be clear. With precautions taken, the tanker moved farther out and anchored for the night.[32]

MINE EXPLODES, DAMAGES GASOLINE TANKER

> *Being high up on the director I could tell the ship already had a sizable list and was down by the bow. As I looked over the side, I could see life jackets, boxes and other gear from the cargo hold floating down the port side. The list was noticeably increasing and more and more objects were coming out of the hold. It looked to me that we could easily be sinking.*

—Paul Gryniewicz describing his view of USS *Noxubee* listing to port, after a mine affixed to the ship, exploded, breeching her hull.[33]

A couple more cases of grenades were broken out, and the fo'c'sle and fantail sentries instructed to toss grenades over the side every few minutes to keep any unwanted company away. Damage control parties remained at their stations, just in case, as those not on watch lay below to get some rest. At 0201, a mine exploded on the ship's port side, forward—producing a column of water that reached the top of the radar mast. Fortunately, the swimmers had not placed it at the stern by the engine room, or the tank deck (where it would have caused the most damage) but instead, near the dry cargo hold.[34]

The blast opened a 3-foot by 5-foot hole in *Noxubee*'s hull, and water surged in, flooding voids beneath the cargo hold and the hold to a depth of two feet. Water also filled the forward magazine to a depth of six feet, but fortunately no fires or personnel casualties resulted from the explosion. The General Quarters alarm sounded; the fo'c'sle deck crew slipped the port anchor chain; and *Noxubee* was under way again. With the port list increasing, the bridge kept announcing distance and bearing to the nearest land, in case it was necessary for the crew to abandon ship. Landing craft coming out from Cua Viet were to serve as rescue craft if necessary.[35]

Such assistance was not needed. Cargo preventing access for vital damage control actions was jettisoned over the side, and repair party personnel stemmed the flooding with mattresses, life jackets, and plywood. Water intrusion persisted, but "plugging the hole" enabled water to be pumped out faster than it was coming in. The bow began to rise, and the list decreased to only a couple of degrees. By 0346, *Noxubee* was no longer in immediate danger of sinking. The salvage ship *Grapple* (ARS-7) was in the area and available to assist the tanker in making emergency repairs.[36]

At first light, *Noxubee* anchored near *Grapple* and secured from General Quarters. Many hours of work by her salvors and the tanker's engineers made *Noxubee* sufficiently seaworthy for the return transit to Da Nang. That evening, she got under way, accompanied by *Grapple*, and limped back to Da Nang Harbor. There, the ship was listed hard over to starboard to allow a steel plate to be welded over the hole. Sufficient other repairs were made that, a week later, she was back delivering cargo to Cua Viet, Tan My, Thon, My Thuy, and Sa Huynh. At month's end, *Noxubee* was released to proceed to Subic Bay for dry docking and permanent repairs.[37]

MOST LAUDED GASOLINE TANKER

Photo 12-13

USS *Patapsco* (AOG-1) lying at anchor off Cua Viet, South Vietnam, 1967. Naval History and Heritage Command photograph #2014.54.01

USS *Patapsco*, under Lt. Larry Smith, USN, was the most decorated of the six gasoline tankers that served in Vietnam. She earned two Combat Action Ribbons, and two Meritorious Unit Commendations for the dates or periods shown in the table.

Gasoline Tanker Unit Awards

Ship	CR	MUC/NUC
Elkhorn (AOG-7)		
Genesee (AOG-8)	22 Apr 1968	NUC: 23 May - 25 Sep 1965
Kishwaukee (AOG-9)	21 Feb 1968	
Noxubee (AOG-56)	28 Oct 1968	MUC: 10 May - 23 Nov 1968
	9 Sep 1969	
Patapsco (AOG-1)	16 Feb 1968	MUC: 1 Oct - 31 Dec 1967
	27-28 Feb 1968	MUC 26 Jan - 31 Mar 1968
Tombigbee (AOG-11)		

13

Stores, Combat Stores, and General Stores-Issue Ships

Photo 13-1

Romeo flags (single letter R signal flags) closed up on the port and starboard yardarms of a replenishment ship communicated to customer ships that she was ready to receive them along both sides of her.
USS *Pictor* (AF-54) South China Sea 1965 cruise book

Two necessities for Navy ships on the line, and vessels inshore in Vietnam, were fuel and ammunition. As such, replenishment alongside oilers (AO) and ammunition ships (AE) was a frequent occurrence. However, diesel fuel, aviation fuel, and gun rounds did not meet the nutrition and other needs of a ship's crew. Accordingly, "floating supermarkets" were also required. These came in the form of general stores-issue ships (AKS), stores ships (AF), and combat stores ships (AFS)—which carried stores, refrigerated items, and some equipment.

With the exception of *Vega*, the stores ships that served in Vietnam were all laid down as merchant ships, or were in service as such from the early to mid-1940s, before the Navy acquired them and converted them for fleet support usage. Four newer purpose-built *Mars*-class combat stores ships deployed later, were longer, broader, and faster—with greater capabilities than the ex-merchantmen.

Stores Ships
Alstede-class: 459 feet; 14,180 tons; 16 kts; 292 ship's complement
Hyades-class: 468 feet; 15,300 tons; 15.5 kts; 22 officers, 230 men
Denebola-class: 455 feet; 11,900 tons; 16 kts; 250 ship's complement
Rigel-class: 502 feet; 15,150 tons; 21 kts; 350 ship's complement

Ship	Comm.	Decom.	Disposition
Aludra (AF-55) launched: 14 Oct 44 *Alstede*-class	19 Jun 52	12 Sep 69	Laid up in the National Defense Reserve Fleet, Suisun Bay, Benecia, CA
Bellatrix (AF-62) launched: 4 Dec 44 *Alstede*-class	18 Nov 61	30 Sep 68	Sold for scrapping, 24 Mar 69, to Zidell Explorations
Graffias (AF-29) laid down in 1943 *Hyades*-class	28 Oct 44	15 Oct 69	Laid up, 20 May 70, in the National Defense Reserve, Suisun Bay, Benecia, CA
Pictor (AF-54) launched 4 Jun 42 *Alstede*-class	13 Sep 50	Dec 69	Laid up in the National Defense Reserve Fleet
Procyon (AF-61) launched 1 Jul 42 *Alstede*-class	24 Nov 61	4 Feb 72	Laid up in the National Defense Reserve Fleet
Regulus (AF-57) launched 7 Jun 44 *Denebola*-class	3 Feb 54	10 Sep 71	Grounded on a reef at Kau yi chau Island, damaged beyond economical repair
Vega (AF-59) *Rigel*-class	10 Nov 55	29 Apr 77	Sold 1 Dec 77, to Union Metals & Alloys
Zelima (AF-49) launched 2 Mar 45 *Alstede*-class	27 Jul 46	Sep 69	Laid up in the National Defense Reserve Fleet, Suisun Bay, Benecia, CA

Mars-class Combat Stores Ships
581 feet; 15,900-18,663 tons; 20 kts; 486 ship's complement

Ship	Comm.	Decom.	Disposition
Mars (AFS-1)	1 Dec 63	1 Feb 93	Placed in service by Military Sealift Command as USNS *Mars* (T-AFS-1)
Niagara Falls (AFS-3)	29 Apr 67	Sep 94	Placed in service as USNS *Niagara Falls* (T-AFS-3)
San Jose (AFS-7)	23 Oct 70	2 Nov 93	Placed in service as USNS *San Jose* (T-AFS-7)
White Plains (AFS-4)	23 Nov 68	17 Apr 95	Laid up in the Naval Inactive Ship Maintenance Facility, Pearl Harbor, HI

GENERAL STORES-ISSUE SHIPS

Photo 13-2

USS *Castor* (AKS-1) with *Pollux* (AKS-4) berthed astern, location and date unknown. *All Hands* magazine, September 1968

Two *Castor*-class general stores-issue ships met other requirements in Vietnam. Slightly older than the other World War II-vintage ships, they also began life as merchant vessels.

Castor-class General Stores-Issue Ships
459 feet; 14,440 tons; 16.5 kts; 15 officers, 184 men

Ship	Comm.	Decom.	Disposition
Castor (AKS-1)	12 Mar 41	30 Jun 47	Sold for scrapping, 25 Jul 69
launched 20 May 39	24 Nov 50	31 Oct 68	to Mitsui & Co., Japan
Pollux (AKS-4)	27 Apr 42	3 Apr 50	Sold for scrapping, 2 Sep 69,
laid down 2 Oct 41	5 Aug 50	31 Dec 68	to Mitsui & Co., Japan

Castor was laid down as the first C-2 class cargo vessel by the Federal Shipbuilding and Dry Dock Company for the U.S. Maritime Commission, and launched at Kearney, New Jersey, on 20 May 1939 as the steamship SS *Challenge*. She was operated by the Cuba Mail Line for one year before her purchase by the Navy on 23 October 1940. Following conversion, *Castor* was commissioned on 12 March 1941, at the Brookland Navy Yard. Her mission was to carry general stores, general store stock, clothing, and medical supplies to the forward operating areas where no advance bases had been established. She earned three battle stars for service in the Pacific Theater in World War II, and two for Korean War service.[1]

Photo 13-3

USS *Castor* moored at berth six, in India Basin, at Sasebo.
USS *Castor* (AKS-1) 1967 cruise book

Castor was home ported at Sasebo, Japan, in 1964. From there, she made regular deployments to Vietnam, in support of ships on Yankee Station or those engaged in MARKET TIME operations, until decommissioned at Sasebo on 31 October 1968. Since early 1966, she resupplied, on average, over one-hundred ships per deployment.[2]

RECEIPT OF REQUIREMENTS TO REPLENISHMENT

Photo 13-4

Boatswain's Mates ready to send the cargo hook down into No. 2 Hold.
USS *Castor* (AKS-1) 1967 cruise book

When a customer ship learned that *Castor* would soon be in their area, the ship's supply officer would create a shopping list in message form and transmit it to the *Castor*. Once Radio Central on board *Castor* received the message, it was sent to the Electronic Accounting Machine Room, which generated punched IBM cards. The cards were sent to the Fleet Issue Office and sorted for distribution to the ship's holds. After the cards reached the holds, sailors began to bag and box the requirements, and placed them on the hatch square. The materials were then ready for breakout.[3]

Breakouts were to *Castor*, as a shopping basket or cart was to a supermarket customer. After picking and packaging was complete, all material was brought topside and positioned on an UnRep (underway replenishment) station, so that it might be delivered to a receiving ship in an orderly fashion.[4]

Castor had eleven topside customer delivery stations—five to port, five to starboard, and a helicopter platform aft. Often the material in a forward hold needed to be staged on a station aft. In such cases, sailors hand carried the stores the length of the ship. These goods, once on station, were placed on wooden pallets and secured in nylon cargo transfer nets. Then the word was passed, "Supply is Ready."[5]

Photo 13-5

Cargo being staged, in preparation for transfer to a ship alongside.
USS *Pictor* (AF-54) South China Sea 1965 cruise book

Photo 13-6

Transfer of materials to customer ship via tensioned span wire and cargo net. USS *Pictor* (AF-54) South China Sea 1965 cruise book

Vietnam Support Provided by USS *Castor* in 1968
(She was decommissioned on 31 October)

Combat Zone	Combat Zone	Combat Zone
Tonkin Gulf: 4-10 Feb	An Thoi: 27 Feb	An Thoi: 9 May
Da Nang run: 11 Feb	Vung Tau: 28 Feb	Vung Tau: 11 May
Cam Ranh run: 12 Feb	Cam Ranh: 29 Feb	Cam Ranh: 12 May
Vung Tau run: 13 Feb	Qui Non: 1 Mar	Da Nang: 13 May
An Thoi run: 14 Feb	Da Nang: 2 Mar	Tonkin Gulf: 14-22 May
	Tonkin Gulf: 4-10 Mar	

STORES SHIP *PICTOR*'S 1965 WESTPAC CRUISE

It was bridge watches and day and night breakout and transfer.... It was six short blasts of the whistle and a close call.... It was moving chow to another station when the rig broke. It was trying to breakout with no conveyer working. It was losing two dozen red night lights on one ship. It was Condition III watches, bush hats and berets.

—From USS *Pictor* (AF-54) South China Sea 1965 cruise book

The primary commodity carried by stores ships such as *Pictor*, was food. Preliminaries to replenishments at sea were similar to those aboard *Castor*. Following receipt aboard ship of the customer's requirements, the total weight of the items ordered was computed and the number of net loads required for the transfer was determined. The day before or the day of the UnRep, depending on the quantity of items, the breakout detail was set. Most of the crew were then involved in getting the food up to the main deck, so that it could be netted for transfer. Immediately before the UnRep, this detail was set once again to breakout the chill and frozen items just before the ship came alongside.[6]

Photo 13-7

Stores ship USS *Pictor*, date and location unknown.
USS *Pictor* (AF-54) South China Sea 1965 cruise book

Meanwhile, the deck crew on each station worked to prepare the transfer rigs that were to be used, and get all the lines, messengers, telephone lines, etc., ready. On the bridge, the best course for the UnRep was determined and this information communicated to the customer. The UnRep officer of the deck, helmsman, and bridge watch were posted, and after steering, and the emergency stations manned. In short, replenishment at sea was something in which everyone had a job to do and a part to play.[7]

Photo 13-8

Pictor's helmsman was responsible for keeping her on a very steady course. This required a constant vigil on the compass, and making minor corrections with the wheel, since even a little deviation in course could result in a collision.
USS *Pictor* (AF-54) South China Sea 1965 cruise book

Many replenishments during *Pictor*'s cruise were done in port. Although her primary mission was underway replenishment, in port supply of ships and shore activities was a valuable service and in many cases, a necessity. The most important ports requiring resupply were Da Nang, Chu Lai, Cam Ranh Bay, Qui Nhon, and An Thoi.[8]

Photo 13-9

Most food transfer in port was via landing craft, due to their load capacity.
USS *Pictor* (AF-54) South China Sea 1965 cruise book

Off Vietnam, *Pictor* was a welcome sight for Marines ashore, whom often had not enjoyed fresh food for a while. When the floating market arrived, word travelled fast and customers lined up for their share. They came rain or shine, enthused about items that seemed commonplace to sailors aboard the stores ship.9

Photo 13-10

Landing craft coming alongside the stores ship *Pictor*.
USS *Pictor* (AF-54) South China Sea 1965 cruise book

Pictor's cruise in 1965 also took her to many ports in Southeast Asia. After a week of liberty in Japan, she headed for Kaohsiung, Taiwan, to transfer food to the combat stores ship *Mars*, which had exhausted much of hers in support of an effort to free a destroyer which had run aground on Pratus Reef. (Details about this operation may be found in the following chapter, devoted to salvage efforts during the war.)[10]

COMBAT STORES SHIP *NIAGARA FALLS*

Photo 13-11

From pieces of steel to a proud ship. Keel laid, 22 May 1965; launched, 26 March 1966; commissioned 29 April 1967; departure on first Western Pacific cruise, 28 March 1968. USS *Niagara Falls* (AFS-3) Western Pacific 1968 cruise book

The seven *Mars*-class combat stores ships (of which four served in Vietnam) were designed for underway replenishment, in support of operating forces. This was accomplished by simultaneously providing refrigerated stores, dry provisions, spare parts, general stores, fleet freight, mail, personnel, and other items from five stations—two on the starboard side and three on the port. By the early 1990s, the class was refitted with limited refuel capacities for F-76 diesel fuel.[11]

Following their naval service in the 1990s, five of the ships—*Mars, Niagara Falls, Concord, San Diego*, and *San Jose*—were taken over by the Military Sealift Command, and continued their service with smaller, less expensive civilian crews. The remaining two were laid up; *Sylvania* was later scrapped and *White Plains* sunk as a target in a naval exercise.[12]

NIAGARA FALLS' MAIDEN CRUISE STATISTICS

Photo 13-12

Banner on bridge wing of USS *Niagara Falls* proudly proclaims number of replenishments she conducted during her 1968 Western Pacific cruise. USS *Niagara Falls* (AFS-3) Western Pacific 1968 cruise book

- Swings (number of times) on Yankee Station – 5
- Swings on Market Time area – 4
- Ships supplied – 407
- Short tons of provisions and general stores issued – 4,950
- Total issues – 72,000
- Money value of sales – 5 million dollars
- Record replenishment: 342 pallets in 1 hour, 56 minutes

14

Early War Salvage Efforts

> *The combat in Vietnam required a considerable amount of direct salvage support both offshore and in that country's harbors, rivers, and canals. Logistic support required operating close to shore, landing through the surf, and steaming through the reefs of the South China Sea. The increased shipping meant more casualties than normal throughout the Pacific, especially since many older ships, subject to frequent breakdown, had been mobilized from the National Defense Reserve Fleet. Operations within Vietnam were subject to casualties from direct combat and from sabotage and sapper operations. Responsibility for the salvage work fell primarily upon the Pacific Fleet.*
>
> —From Charles A. Bartholomew, and William I. Milwee Jr.'s, authoritative book, Mud, Muscle, and Miracles: Marine Salvage in the United States Navy.[1]

Salvage in support of combat operations placed heavy demands on Pacific Fleet salvage force resources during the Vietnam War. Eight salvage ships (ARS) and eighteen fleet tugs (ATF)—all of World War II construction—formed the initial force at war's commencement. Several auxiliary tugs (ATA) were also in service, but they offered only very limited salvage capabilities.[2]

Fortunately, the ARSs, considerably upgraded since World War II, boasted highly reliable Caterpillar diesel engines, and carried eight legs of beach gear. New diesel-driven Power Pack salvage machinery had begun replacing vintage, gasoline-driven machinery, and the ships' diving capability had also been upgraded. Each ARS was allocated a complement of eighteen divers. "Mustang" officers—ex-enlisted men who had come up through the ranks—commanded about half the ships, and young regular officers, who exhibited great potential to ascend to senior ranks, the remainder of them.[3]

The Navy's Supervisor of Salvage was Comdr. Willard Searle Jr., who'd been replaced in his previous job, Pacific Fleet Salvage Officer, by Comdr. Eugene B. Mitchell. These two men made a powerful team, anticipating the salvage needs in Vietnam, and taking action to meet them. One important capability was missing from the Pacific Salvage Fleet: a mobile salvage unit that could provide rapid salvage assistance in the rivers, harbors, and canals of Vietnam. To meet this need, Mitchell and Searle worked to create and outfit a new organization.[4]

Harbor Clearance Unit One (HCU-1) was established at Subic Bay on 1 February 1966, with a hand-picked cadre of five officers and sixty-five men comprising its complement of personnel. Because it was a Western Pacific unit, HCU-1 would normally have come under the control of commander, Service Group Three. Instead, the Service Force commander retained operational control of the unit, except for teams deployed to Vietnam—thereby ensuring flexibility of it to operate both in Vietnam and throughout the Pacific.[5]

The remainder of this chapter focuses on salvage ship operations efforts off Vietnam, and elsewhere in the Pacific, during the early part of the war. (Summary information about accomplishments of HCU-1 warranting Meritorious Unit Citations may be found in Appendix C.)

KNOX ON THE ROCKS

Photo 14-1

Destroyer USS *Frank Knox* (DDR-742) aground on Pratas Reef, South China Sea, with several ships attempting to pull her off. They are (from left to right): *Grapple* (ARS-7), *Conserver* (ARS-39), *Sioux* (ATF-75), *Greenlet* (ASR-10) and *Cocopa* (ATF-101).
Naval History and Heritage Command photograph #NH 74179

In early morning darkness on 18 July 1965, while steaming at 16 knots in the South China Sea, the destroyer *Frank Knox* (DDR-742) went hard aground at 0235 on Pratas Reef, some two hundred miles east-southeast of Hong Kong. A survey by Taiwanese UDT divers found the destroyer was aground for about half her length, and had suffered some hull and propeller damage.[6]

The salvage ship *Grapple* (ARS-7), and fleet tugs *Munsee* (ATF-107) and *Cocopa* (ATF-101), already at sea, were ordered to the scene. Separately, the Seventh Fleet Salvage Officer, Lt. Comdr. John Huntly Boyd Jr., left Sasebo for Pratas Reef. The salvage effort would ultimately also involve the salvage ship *Conserver* (ARS-39), tug *Sioux* (ATF-75), and submarine rescue ship *Greenlet* (ASR-10). When the first ships arrived off the reef-gripped *Knox*, their initial priorities were to remove weight from the destroyer, and rig a beach gear leg aboard her to prevent broaching, before attempting to pull her free.[7]

Photo 14-2

USS *Frank Knox* (DDR-742) aground on Pratas Reef, in July 1965. A H-34 helicopter is hovering over her bow to evacuate crewmen. This was the only safe method of transportation to and from the stranded ship during the rough seas that persisted during most of the several weeks of salvage operations that finally freed her.
Official U.S. Navy photograph #USN 1113134

Grapple had Army barges in tow when ordered to the scene. One was placed alongside *Knox* to receive ammunition. In a related action, preparatory to attempting to refloat her, ship's engineers pumped fuel aft to lighten the destroyer forward. Meanwhile, *Munsee* and *Grapple* began laying beach gear. Some difficulties were encountered and, when the tide was right for the pull at midday on 20 July, only the leg to *Knox* and one to *Grapple* were ready. Nevertheless, the attempt was made. After three hours, the destroyer had shifted about twelve feet. Success seemed at hand. Additional beach gear would be available for the next attempt to free her, at high tide the following day.[8]

Such was not to be. Typhoon GILDA, well to the south and expected to dissipate while passing over the Philippines, regenerated later that day after clearing Luzon. Following the initial pulling effort, the westerly winds shifted to the southeast, the seas began to build rapidly, and *Knox* took a beating. She began to roll and work against the reef, and her head swung to starboard. *Grapple* hauled on the beach gear, and went full ahead in an effort to prevent the destroyer from broaching. But the beach gear anchor dragged. With the loss of this fastening point (and associated ability to assist the destroyer), and with the weather worsening, *Grapple* slipped her gear and cleared the area.[9]

It appeared prudent for the destroyer's crew to abandon, but high seas made it impossible to do so by boat. Commander, Seventh Fleet ordered the aircraft carrier *Midway* (CVA-41) to the scene, as well as an amphibious task force that included *Iwo Jima* (LPH-2), so that their helicopters could help with the abandonment. By midnight that same day, the 20th, the worse of the storm had passed, eliminating the need for *Knox*'s crew to leave. But the seas had moved the destroyer laterally about seventy-five feet. She was now entrenched in the coral, with openings in her hull and forward machinery spaces flooded.[10]

The revised salvage plan to refloat *Knox* called for dewatering her, and increasing pulling force through the use of additional ships. For such, the salvage ship *Conserver* (ARS-39), fleet tug *Sioux* (ATF-75), and submarine rescue ship *Greenlet* (ASR-10) were ordered to the scene. As the destroyer continued to grind against the reef, she developed additional leaks in her bottom. Pumps and compressors were brought on board to dewater flooded spaces, and to blow fuel tanks (empty them of sea water) in preparation for an attempt to free her on 25 July.[11]

This attempt failed and, with Typhoon HARRIET approaching, additional pumps were brought aboard, procedures for blowing tanks and compartments refined, patching efforts continued, and beach gear laid to enable *Cocopa* to pull from directly astern of *Knox*. The pull was made on the 26th, on rising seas, again without success. *Grapple*'s tow wire and *Cocopa*'s tow winch failed, leaving *Conserver* heaving and *Frank Knox* backing hard, and the destroyer turning in the wrong direction. She moved about thirty feet but still clung hard to Pratas Reef—where she had now been for eight days.[12]

TECHNOLOGY/PROVEN METHODS YIELD SUCCESS

Ensuing efforts to free *Knox* on 31 July and 2 August were unsuccessful. She was now more severely holed, with weakened hull structure, and machinery spaces still flooded. Conventional hole patching and water removal methods having proved inadequate, new methods were tried. The salvors filled the flooded compartments with plastic foam, which expelled the water and also greatly enhanced the destroyer's buoyancy. Hull stiffeners were welded to the main deck, and explosives used to break up coral around the ship. The latter action caused further damage, necessitating the use of more foam. Another pull was attempted on 11 August, which failed as well.[13]

Over the next several days, salvage tackle was re-rigged, additional weight was removed from *Knox*, pontoons were attached to her hull, and additional foam was generated. Also, the destroyer *Cogswell* (DD-651) arrived to make waves to help break the reef's grip on the grounded destroyer. This was done by steaming close aboard, and generating the desired wake. A pulling effort on 22 August moved *Frank Knox* eight feet aft. On 24 August, she was finally afloat, after nearly six weeks of salvage work in a very difficult environment. Through much effort, using the latest in technology, coupled with tried-and-true salvage methods, she had been saved. The destroyer was repaired in Japan, and gave the U.S. Navy another six years of service, followed by two decades in the Greek Navy as *Themistoklis* (D210).[14]

THIRTY-FOUR SALVAGE SHIPS

The salvage vessels involved with saving the *Frank Knox* were part of a larger group of thirty-four—9 salvage ships, 17 fleet ocean tugs, 4 submarine rescue ships, and 4 auxiliary fleet tugs—that served in Vietnam. These ships are identified in the tables along with the Combat Action Ribbons, Meritorious Unit Commendations, and Navy Unit Commendations they earned during the war.

Nine Salvage Ships (ARS)
Bolster-class: 213 feet; 2,048 tons; 16 kts; 7 officers, 113 men
Diver-class: 213 feet; 1,897 tons; 15 kts; 7 officers, 113 men

Ship	CR	MUC/NUC
Bolster (ARS-38)	7 Feb 68	
Bolster-class	16 Feb 68	
	21 Feb 68	
	4-5 May 70	

Conserver (ARS-39) *Bolster*-class		MUC: 30 Nov 70-4 Jan 71
Current (ARS-22) *Diver*-class		MUC: 27 Jan-20 Jul 67
Deliver (ARS-23) *Diver*-class	27-28 Feb 68	MUC: 13-27 Apr 75 MUC: 1-7 May 75
Grapple (ARS-7) *Diver*-class		
Grasp (ARS-24) *Diver*-class		MUC: 5 Jul-1 Sep 68 MUC: 16 Feb-7 Apr 69 MUC: 4 Jan 70-29 Mar 72
Opportune (ARS-41) *Bolster*-class		MUC: 4 Apr 68 MUC: 26 May-11 Oct 69 MUC: 28 May-22 Jul 75
Reclaimer (ARS-42) *Bolster*-class		MUC: 5 Feb-10 Aug 72
Safeguard (ARS-25) *Diver*-class		MUC: 17-18 Mar 73

Seventeen Fleet Ocean Tugs (ATF)
Abnaki-class (***Abnaki, Arikara, Chowanoc, Cocopa, Hitchiti, Moctobi, Molala, Munsee, Shakori, Tawakoni***):
205 feet; 1,675 tons; 16.5 kts; 5 officers, 80 men
Navajo-class (***Apache, Lipan, Mataco, Quapaw, Sioux, Tawasa, Ute***):
205 feet; 1,675 tons; 16.5 kts; 5 officers, 80 men

Ship	**MUC**	**NUC**
Abnaki (ATF-96)	1 Sep-2 Oct 69 13-27 Apr 75	
Apache (ATF-67)	1 Sep 71-23 May 72	3 Feb-7 Oct 69
Arikara (ATF-98)		
Chowanoc (ATF-100)		
Cocopa (ATF-101)		
Hitchiti (ATF-103)	2 Sep-15 Nov 69	
Lipan (ATF-85)	2 Oct-5 Nov 69	
Mataco (ATF-86)		
Moctobi (ATF-105)		
Molala (ATF-106)		
Munsee (ATF-107)		
Quapaw (ATF-110)	13-27 Apr 75	
Shakori (ATF-162)		
Sioux (ATF-75)		
Tawakoni (ATF-114)	16 Aug-9 Sep 71	
Tawasa (ATF-92)	16 Aug-9 Oct 71	
Ute (ATF-76)	27 Aug-15 Nov 69	

Four Submarine Rescue Vessels (ASR)
Chanticleer-class: 251 feet; 2,141 tons; 14 kts; 6 officers, 96 men

Ship	MUC	NUC
Chanticleer (ASR-7)		
Coucal (ASR-8)	4-18 Mar 70	
Florikan (ASR-9)		
Greenlet (ASR-10)	11 Oct 69-14 Apr 70	

Four Auxiliary Fleet Tugs (ATA)
Sotoyomo-class: 143 feet; 1,360 tons; 12 kts; 5 officers, 40 men

Ship	MUC	NUC
Mahopac (ATA-196)		
Sunnadin (ATA-197)		
Tillamook (ATA-192)	1 Dec 66-1 Nov 68	
Wandank (ATA-204)		

Combat in Vietnam created constant requirements for salvage work, beginning in 1965, which continued for several years. Overviews of a few representative salvage operations through October 1967 follow. They involved refloating ships mined by the enemy, or which had grounded in Vietnamese waters. Salvage operations in Vietnam and elsewhere in the Pacific included other vessels, as well as aircraft, ordnance, and other things of sufficient value to retrieve from the sea.[15]

TERRELL COUNTY HARD AGROUND AT TUY HOA

Photo 14-3

USS *Terrell County* (LST-1157) under way, location and date unknown.
Navy photo courtesy of Capt. John L. Townley USN (Retired), and NavSource

Salvage forces were called upon in late November that same year, 1965, to pull the tank landing ship *Terrell County* (LST-1157) off the beach at Tuy Hoa. While delivering Republic of Korea Marines and U.S. Army support units to the coastal town midway between Nha Trang and Qui Nhon, and with the troops still aboard, she had broached in heavy surf and went aground.[16]

Arriving on the scene, fleet tug *Molala* (ATF-106) laid one leg of beach gear and, pulling in tandem with the auxiliary tug *Mahopac* (ATA-196), hauled the LST off the beach on 24 November. Subsequently patched and pumped dry, *Terrell County* got under way on 2 December, under tow for Yokosuka. Permanent repairs to her hull were completed on 22 February 1966, and she left for Naha, Okinawa.[17]

Photo 14-4

Fleet auxiliary tug USS *Mahopac* (ATA-196) under way, location and date unknown. Naval History and Heritage Command photograph #L45-173.01.01

TANKER *SEA RAVEN* AND TANK LANDING SHIP *SUMMIT COUNTY* BOTH GROUND IN CHU LAI AREA

On 23 February 1966, the small Panamanian-registry T-1 commercial tanker *Sea Raven* went aground about one hundred yards from the beach in the vicinity of Chu Lai. As tank landing ship *Outagamie County* (LST-1073) stood by, bad weather and heavy seas made it difficult to render assistance. The crew was taken off by helicopter before dark and by rubber boat during the night.[18]

Hitchiti (ATF-103) arrived on the scene the next day, joined on the 26th by *Reclaimer* (ARS-42) for salvage operations after the seas abated sufficiently for salvage personnel to board the stricken ship and survey her damage. *Bolster* (ARS-38) arrived on 1 March and relieved *Hitchiti*. Work by the salvage ships, their crews, and a contingent from the newly formed Harbor Clearance Unit One, progressed rapidly. De-watering and removal of cargo was completed on 3 March and, two days later, *Bolster* and *Reclaimer* succeeded in refloating *Sea Raven*. *Reclaimer* then took her under tow to Cam Ranh Bay, where she was transferred to a commercial tug on 8 March 1966.[19]

Photo 14-5

USS *Hitchiti* (ATF-103) under way off the coast of Hawaii, 12 January 1970. National Archives photograph #K-81123

Three days later, the tank landing ship *Summit County* (LST-1146) grounded near Chu Lai, resulting in flooding in her engine room. (Survey ship operations to sound the sea approaches to Chu Lai, and produce charts necessary for ships to safely navigate, did not commence until mid-March.) *Hitchiti* arrived on 12 March to assist ship's force in patching the engine room breech and preparing the ship for tow. Once *Summit County* was seaworthy, she took her in tow, bound for Sasebo and permanent repairs. On 24 March, *Bolster* took over this chore near Taiwan, to allow the tug to visit Hong Kong for a well-deserved period of rest and recreation.[20]

BRITISH TANKER *AMASTRA* MINED AT NHA TRANG

> *At approximately 0015 the whole world seemed to turn upside down. A massive explosion awakened me and simultaneously the ship felt as though it was leaping out of the water and a huge searing blue flash from outside lit up my bedroom. She settled back into the water, rocking and groaning and with the noise of the shockwave echoing all around.*
>
> —Colin Avery, second engineer aboard MS *Amastra*, describing an explosion on 12 April 1967, caused by the detonation of swimmer-sapper emplaced explosives on the hull of the Shell Oil tanker.[21]

Shortly after midnight, in pitch-black on the morning of 12 April 1967, a terrific explosion pierced the quiet of the harbor at Nha Trang. As dawn broke, it was possible to see the MS *Amastra*, a British-registered Shell Oil tanker, sunk by the stern, resting in about 60 feet of water.[22]

The motor ship, under the command of Capt. John Campkin, had begun unloading 15,000 tons of aviation fuel into an underwater pipeline the previous day. Fortunately, most of the JP4 fuel had been discharged by the time of the detonation. The explosives placed on the hull by Viet Cong frogmen, opened a 6-foot by 4-foot hole, about ten feet below the waterline. Fortunately, there was no loss of life, or serious injury to any of her forty-three-man crew.[23]

As the volume of water entering the hull overtook the efforts of the ship's pumps, *Amastra* gradually settled by the stern. Most of the crew left her at about 0400, and were taken to the 5th Special Forces compound in Nha Trang, and to nearby Camp McDermott.[24]

Current (ARS-22) arrived on 13 April, and put into motion necessary actions to raise the *Amastra*. Over the next several days, fuel trucks lined the beach area. They worked in concert with the Dutch-flagged *Kara*, to offload 640,000 gallons of fuel from the tanker, to lighten her for refloating. Within days, *Greenlet* (ASR-10) was also ordered to the scene. After *Current*'s crew fabricated and installed a patch over the gaping hole in *Amastra*'s hull, pumps were started, water began returning to the sea from her compartments, and she rose to the surface.[25]

Once sufficiently seaworthy, *Amastra* departed for Singapore on 29 April, under tow by two commercial tugs. She entered the dockyard there in May. Twenty months later, another Shell tanker, MS *Helisoma*, was mined and sunk in Nha Trang Harbor on 22 December 1968. This occurred in nearly the same spot and with, once again, no casualties.

Viet Cong from the nearby village of Truong Tay were believed responsible for both mining incidents.²⁶

COCONINO COUNTY MINED AT CUA VIET

Photo 14-6

USS *Coconino County* (LST-603), with a helicopter on deck, during landing operations at the Co Chien River mouth, as part of Operation DECKHOUSE V, 6 January 1967. National Archives photograph #K-35546

On 29 June 1967, the tank landing ship *Coconino County* (LST-603) was mined while beached on the LST ramp at Cua Viet, discharging cargo. The explosion critically wounded one of her crew, and tore a three- by nine-foot hole in the ship's bottom—flooding a generator room and the main and auxiliary engine rooms. Three hours later, a second mine exploded about fifteen feet off *Coconino County*'s starboard quarter. The blast loosened the ship's stern-tube packing, resulting in some flooding of the port and starboard shaft alleys.²⁷

This attack by swimmer-sappers was yet another in a series of attempts by the enemy to disrupt waterborne traffic from Cua Viet (located at the mouth of the same-named tributary), upriver to the U.S. Marine Corps base at Dong Ha. Cua Viet was the Navy base farthest north in Vietnam. As such, it was under mortar, rocket, and ground attack by the North Vietnamese Army for most of the war. The Da Nang Naval Support Detachment at Cua Viet had been established in March 1967, to augment the efforts of the nearby base at Dong Ha. Cua

Viet served as a trans-shipment point for supplies headed for Dong Ha.[28]

Life at Cua Viet was made difficult both by enemy fire and the physical environment. Winds and rains of the winter monsoons were harsh and, outside the river mouth, shoals endangered ships making the run up the coast from Da Nang. Crossing the bar involved shifting sand bars, which required continuous dredging, and periodic enemy artillery fire. *Caroline County* (LST-525) had become the first major naval vessel to enter the river and berth at Cua Viet, when she did so in March. *Snohomish County* (LST-1126) was the second ocean-going visitor to Cua Viet. Together, Cua Viet and Dong Ha provided logistical support to American and allied units operating in I-Corps around the DMZ area.[29]

Hitchiti took *Coconino County* under tow on the same day of the mine explosions, and delivered her to Da Nang. This coastal passage followed the removal of all cargo from the ship, and installation of temporary patches over the damaged areas of her hull.[30]

TRANSPORT *GEIGER* GROUNDS AT DA NANG

Photo 14-7

Transport USNS *Geiger* arriving at the naval base at Rota, Spain, 17 May 1959. Naval History and Heritage Command photograph #NH 92360

On 4 October 1967, the transport USNS *Geiger* (T-AP 197) with 1,700 Republic of Korea troops on board went hard aground in thirteen feet of water in Da Nang Harbor. After removing all excess weight from the ship, six harbor tugs and three commercial tugs, working together, attempted to pull her free on the evening of 5 October. All efforts failed to dislodge the transport.[31]

The following morning, the army dredge *Davidson* began removing sand and mud from the port side of *Geiger*, and the area around her bow. Shortly after midnight on the 7th, she was refloated with assistance from the fleet tug *Abnaki* (ATF-93) and several harbor tugs. A subsequent underwater hull inspection revealed no damage.[32]

Photo 14-8

Fleet tug USS *Abnaki* (ATF-96) keeping the Soviet trawler *Gidrofon* under surveillance in the South China Sea, December 1967.
National Archives photograph #K-43379

When not assigned salvage work, fleet tugs and salvage ships were often assigned "blocking ship" duty, which involved shouldering Soviet intelligence-gathering trawlers away from American aircraft carriers. Trawlers near carrier task force operations often attempted to disrupt flight operations by causing the carrier to maneuver radically. A salvage vessel countering such harassment, positioned herself between the trawler and carrier in such a way as to become the "privileged" (stand on) ship under the International Rules of the Road. The "burdened" (give way) trawler would then have to maneuver clear. This required sailing in close company and, often, steely nerves.[33]

15

Royal Australian Navy Clearance Diving Team 3

Photo 15-1

Painting *Ambush on the Long Tau* by Richard DeRosset, the cover art for the author's *Wooden Ships and Iron Men, Volume III: The U.S. Navy's Coastal and Inshore Minesweepers, and the Minecraft that Served in Vietnam, 1953-1976*, published in 2011.

On the Gunline, Gators Offshore and Upriver, and *Support for the Fleet* describe the service and sacrifices of the crews of U.S. Navy ships that operated off the coast of Vietnam, and in dangerous inland waterways during the war in Southeast Asia. These books also highlight the valiant contributions of the Royal Australian Navy to the war effort. Australia was the only allied nation to provide naval support to the United States during the Vietnam War. This support consisted of (1) four destroyers; (2) a detachment of RAN helicopters and aircrews (Royal Australian Navy Helicopter Flight Vietnam); (3) Clearance Diving Team 3; and (4) three logistics support/troop transport ships which carried soldiers and equipment of Royal Australian Regiment (RAR) between Australia and Vietnam. The service of these vessels—HMAS *Sydney*, HMAS *Boonaroo*, and HMAS *Jeparit*—is taken up in Chapters 4-5 of this book.[1]

The RAN warships—HMAS *Brisbane*, HMAS *Hobart*, HMAS *Perth*, and HMAS *Vendetta*—provided naval gunfire support to allied ground forces in South Vietnam and participated in SEA DRAGON operations off North Vietnam. The latter involved runs in to within five miles of a hostile shore, to strike Vinh, Haiphong, and other targets, which often preceded duels with shore batteries. Most such action occurred at mission completion as ships zigzagged, while racing seaward at high speed to clear the coast, to throw off the aim of enemy gunners.[2]

Much of the service of CDT 3, as detailed in *Gators Offshore and Upriver*, involved the inspection of anchor chains, rudders, and propellers of merchant and naval shipping at Vung Tau. These measures were carried out daily for the defense of shipping against enemy attack, while hull searches were performed on request, or following a report of a suspected swimmer in an anchorage. The divers' heroic removal and rendering harmless of deadly ordnance they found, emplaced by enemy swimmer-sappers, saved many ships and lives. During their time at Vung Tau, CDT 3 members searched 7,573 ships, removing 78 explosive devices.[3]

DIVERSITY OF RAN CLEARANCE DIVING TEAM 3

It is with sincere pleasure I commend the Commanding Officer and team members of Her Majesty's Navy Clearance Diving Team Three for service during the period 18 February 1967 to 27 August 1967.

You were assigned to Task Unit 115.3.4 with the primary duty of providing explosive ordnance detection and disposal capability in one of the major shipping areas in Vietnam. Over and above the exacting task of your primary assignment, you were called upon for numerous missions working with Coastal Groups of the Vietnamese Navy, Coastal Surveillance Forces of the US Navy and other forces who required your unique services. Your consistently high state of readiness and the cooperative spirit exhibited by you and your team in all operations contributed much to the success of your efforts against a tough and determined enemy.

The professional competence of your team has made your presence as an integral part of this Task Group a most rewarding experience. All of us with whom you have worked so closely hope we may serve together again in the future. You have done credit to your Navy and yourselves and you have been an inspiration to those of us with whom you have worked. Well done.

—Letter of Commendation from Comdr. William C. Nation USN, Commander Southern Surveillance Group (CTG 115.3), 15 August 1967, to Commanding Officer CDT 3.

In addition to duties previously described, as the U.S. command became increasingly aware of the versatility and positive 'can do' mindset of the Australian clearance divers, they were regularly employed as the 'immediate response team' to any EOD/diving/salvage incident which was 'outside the square.' Descriptions of independent and joint operations by CDT 3 with Seventh Fleet salvage forces, are not included in *Gators Offshore and Upriver*. This chapter covers two such, involving the salvage of the minesweeping boat *MSB-49* and materiel from the Army craft *YFU-63* by personnel of Contingents One and Eight.

The identity of Contingent personnel follow:

First Contingent (6 February – 25 August 1967)	
Lt. Michael T. E. Shotter, RAN	ABCD Philip C. Kember, RAN
POCD Brian V. Clark, RAN	ABCD Geoffrey D. Lassau, RAN
LSCD Peter Boettcher, RAN	ABCD Peter A. Magnuson, RAN

Eighth Contingent (29 October 1970 – 19 April 1971)	
Lt. Edward W. Linton, BEM RAN	ABCD Larry J. Digney, RAN
CPOCD John J. Gilchrist, RAN	ABCD Anthony L. Ey, RAN
POCD Phillip C. Narramore, RAN	ABCD Brian J. Furner, RAN

CPOCD: Chief Petty Officer Clearance Diver
LSCD: Leading Seaman Clearance Diver
POCD: Petty Officer Clearance Diver
ABCD: Able Seaman Clearance Diver

MOST DANGEROUS WATERS

We know what has to be done and we'll do it.

—Statement by BM1 John O. Hood, boat captain of *MSB-45*, to a journalist, following the loss of his 57-foot minesweeping boat to a Viet Cong command-detonated mine in the Long Tau Channel. The explosion had killed one member of the six-man crew and wounded four of the five survivors.[4]

Shortly after the CDT First Contingent arrived in Vietnam, it was assigned to a U.S. Navy EOD (Explosive Ordnance Disposal) team in Saigon, and split in two. The second team—Leading Seaman Peter Boettcher, and Able Seamen Phil Kember and Peter Magnuson—were tasked with the underwater repair and salvage of the minesweeping boat *MSB-49*, ambushed by Viet Cong on 15 February 1967. A portion of the opening chapter of my book, *Wooden Ships and Iron Men, Volume III: The U.S. Navy's Coastal and Inshore Minesweepers, and the Minecraft That Served in Vietnam, 1953-1976*, is devoted to this incident and the heroic actions of *MSB-51* in coming to her aid. Richard DeRosset's painting *Ambush on the Long Tau*, depicting the exchange of gunfire between the lead boat and well-fortified enemy positions on both banks of the channel, is the cover art for the book.[5]

Early that morning, minesweeping boat *49* had backed clear of her berth in a small nest of craft at Nha Be, twisted about and headed downriver to the junction of the Soi Rap River and the Long Tau Channel, taking the left fork into the channel. Following her was *MSB-51*. After entering the channel, the two boats streamed chain-drag sweep gear astern as they proceeded downstream. Comprised of a length of stout chain to which metal spikes were welded every few links, the chain was designed, when dragged across the river bottom, to sever electrical wires used by the Viet Cong to command-detonate mines.[6]

Photo 15-2

Minesweeping boat *MSB-49* operating on the Saigon River in July 1966. National Archives photograph #USN 1118586

At 0655, as the boats made a starboard turn around the first large bend in the Long Tau some five miles downstream from Nha Be, they sailed into a Viet Cong ambush. The lead boat, *MSB-49*, was engulfed by crisscrossed 75mm recoilless rifle and heavy automatic weapons fire. Three rifle rounds pierced her port side, one setting the fuel tanks ablaze. The *MSB-51*, as well as river patrol boats (PBRs) in the vicinity, responded with counter-fire. With the *MSB-49* rapidly taking on water, BM1 John Hood brought the *MSB-51* in close behind her and, with the assistance of a PBR, pushed the besieged craft aground and moored it to trees along the bank.[7]

While PBRs helped keep the enemy pinned down, Hood's crew evacuated the wounded from the burning boat and stripped it of armament. For his actions that day, Hood received the Silver Star Medal. The Navy craft continued to exchange fire with the enemy until 0710, when help arrived from Nha Be in the form of a Navy "Seawolf" helicopter fire team from HC-1 Detachment Twenty-seven. This unit, charged with protecting the Rung Sat, flew ex-Army UH-1Bs.[8]

When the CDT 3 salvage team arrived on scene, they found *MSB-49* abandoned and beached on the bank; her entire crew, all wounded, having been evacuated. There was then still fighting ashore in the near vicinity and rockets flying over the stricken minesweeper. Nevertheless, Kember and Magnusson quickly donned diving gear and set to work. They patched a gaping hole beneath her waterline, using an engine room access cover, and passed lines under her hull to lash it in place. This allowed a salvage boat working with them to pull the minesweeper off the bank and tow it back to Nha Be.[9]

Rear Adm. Norvell G. Ward, commander Naval Forces, Vietnam, sent the following congratulatory message, regarding the operation:

> Successful salvage of *MSB 49* by salvage party composed of COMNAVFORV [Commander Naval Forces, Vietnam] Salvage Officer, and personnel from MINRON DET 11A [Mine Squadron 11 Detachment Alfa], NAVSUPPACT [Naval Support Activity] Nha Be, COMNAVFORV EOD Team and Royal Australian Navy EOD Team is noted with pleasure.
>
> I extend my personal congratulations and appreciation to those officers and men who accomplished this task.

"BAD DAY AT BLACK ROCK"

February 15th, which was marked by ambush after ambush on the river, was later termed by MSB sailors the "Bad Day at Black Rock," after the 1955 Hollywood film of the same name staring Spencer Tracy. In late morning, *MSB-45*, operating in company with *MSB-22* fifteen miles southeast of Nha Be, was lost to a controlled-mine near the west bank of the Long Tau, sinking almost immediately after a violent explosion. Fortunately, the *22* boat was able to recover five crewmen from the water, four of them wounded. A helicopter evacuated the injured to Nha Be and a search began for the missing crewman, Damage Controlman Third Gary C. Paddock, whose body was found three days later. Divers subsequently stripped the unsalvageable boat, which was lying on the bottom, and destroyed the hulk with explosives.[10]

Mid-afternoon that same day, *MSB-51* again came under attack. Hit in the stack and sweep winch by heavy weapons fire, she, along with *MSB-32*, and their two PBR escorts, reversed course and headed north. Two miles upstream, they were ambushed again. Four more PBRs joined the action, accompanied by helicopter and fixed-wing airstrikes on the enemy, in what one participant would later describe as "one hell of a shootout."[11]

The total losses suffered by Mine Squadron 11 Detachment Alfa during four separate actions that day were destruction of *MSB-45*, heavy damage to *49* (which, being unsalvageable, was subsequently destroyed), and damage to *51*, with two sailors killed and sixteen wounded.[12]

SALVAGE OPERATIONS ON ARMY CRAFT *YFU-63*

Photo 15-3

An Australian Army Centurion tank of the First Armoured Regiment (1AR) being unloaded from the U.S. Army craft *YFU-63*, on the outskirts of Baria, the capital of the Phuoc Tuy Province, 12 September 1968.
Australian War Memorial photograph No. EKT/68/0173/VN

On 3 November 1970, U.S. aircraft en route to Da Nang, sighted a harbor utility craft overturned and grounded on a stretch of beach east of Hue. The 115-foot Army *YFU-63* had been transporting 150 tons of ammunition (white phosphorus, 105mm and 81mm rounds) from Da Nang to Tan My, a port facility near Hue, when she capsized in heavy seas. (The vessel was the former U.S. Navy tank landing craft *LCT(6)-989*, which had taken part in the capture and occupation of Saipan and Guam during World War II.) All twelve members of her crew perished. The remains of one member was later found washed ashore on an island to the south of Da Nang. The other eleven were unaccounted for.[13]

Salvage operations began two days later, led by Lt. "Jake" Linton, RAN, who was assigned as the on-scene commander. Supporting the operation was a multi-faceted joint service team comprised of U.S. Army helicopters, mechanized heavy equipment, and perimeter security forces; joined by the salvage ships USS *Grapple* (ARS-7) and USS *Cohoes* (ANL-78), the Philippine tug *Trojan*, and U.S. and Australian divers. Edward W. Linton was officer in charge of the Eighth Contingent, the last sub-unit of Clearance Diving Team 3 to deploy to Vietnam.[14]

Photo 15-4

Royal Australian Navy Clearance Diving Team 3 (8th Contingent). L – R" ABCDs Brian Furner and Anthony Ey, Lt. Edward "Jake" Linton, CPOCD John Gilchrist, ABCD Larry Digney and POCD Phillip Narramore.
Jake Linton Collection

(Linton had received the British Empire Medal in 1962 for his participation in extremely hazardous diving operations at great depths "to clear a blockage in the mouth of the main outlet from [the] Lake Eucumbene [Dam]." Requiring dives to 285 feet, and recognizing that only Navy clearance divers were capable of descending to such extreme depths, the Snowy Mountains Authority requested assistance from the RAN.)[15]

Despite adverse weather and surf conditions, the participants in the operation felt a sense of urgency, owing to the then unknown fate of any crewmembers who might be aboard, and the possible presence of a cargo of high explosives. Battling rough surf conditions generated by Typhoon PATSY, the divers eventually were able to attach tow lines to *YFU-63*, which Navy tugs offshore took a strain on in an attempt to overturn the craft. After repeated efforts failed, bulldozers were employed in an effort to drag the largely submerged craft up onto the beach. This too proved unsuccessful, and a decision was made to build a ramp of sand from the beach out to the wreck to gain access to the hull of the inverted craft.[16]

Photo 15-5

Members of CDT 3 (8th Contingent) attaching a jackstay to the overturned Army craft. Jake Linton collection

With heavy seas breaking over the wreck, an access hole was cut into the hull using oxygen acetylene torches and pneumatic chisels. A search of the internal compartments of the craft found no crewmen.

Divers were able to ascertain the cargo had dispersed, and retrieve important documents and logs. Recovery operations were terminated on 22 November due to deteriorating weather conditions.[17]

Capt. Maurice A. Horn, USN, commanding officer, Naval Support Activity Da Nang, recommended that Lieutenant Linton be awarded the Navy Commendation Medal (with "V" device to denote an act of valor). In the closing paragraph of his justification for the award, Horn wrote:

> Lt. Linton was able to provide that rare quality of responsibility and devotion to duty which was required to lead his men in working 18-hour days and sleeping in austere self-constructed shelter on an exposed beach during the height of the monsoon and typhoon season under the constant threat of enemy attack. His exceptional professionalism, exemplary leadership and devotion to duty were outstanding and brought great credit to himself, his salvage team, this command and his "Service."

No further action was taken as it was Australian Government policy that military members could not wear medals awarded by foreign governments. As such, the prevailing attitude appeared to be that "if they can't be worn, there doesn't seem to be any point in receiving such awards." Chief Petty Officer Gilchrist, Petty Officer Narramore, and Able Seamen Ey and Furner did receive individual letters of commendation. (A copy of Ey's letter may be found in Appendix D.)[18]

CLEARANCE DIVING TEAM 3 DEPARTS VIETNAM

Photo 15-6

Main Gate of the Vung Tau Harbor Entrance Control Post, circa 1968.
Courtesy of Albert Moore, Mobile Riverine Force Association

In April 1971, CDT 3 was relieved of its responsibilities and the Eighth Contingent departed Da Nang and left Saigon for Australia on 5 May. It was the end of CD involvement in the war, most spent at Vung Tau. The First Contingent had been initially employed in Saigon with the USN EOD team to settle in and undergo local indoctrination courses. At the end of February 1967, it had relocated to Vung Tau and moved into one of the bunkers at the Harbor Entrance Control Post.[19]

Subsequent contingents served at Vung Tau until August 1970 when, after being relieved at Vung Tau by South Vietnamese Navy personnel, the Seventh Contingent and their equipment were airlifted to the northern city of Da Nang. The eighth and final contingent served its entire tour based out of Camp Tien Sha in Da Nang, assuming responsibility for Naval EOD in the whole of Military Region I up to the Demilitarized Zone.[20]

Photo 15-7

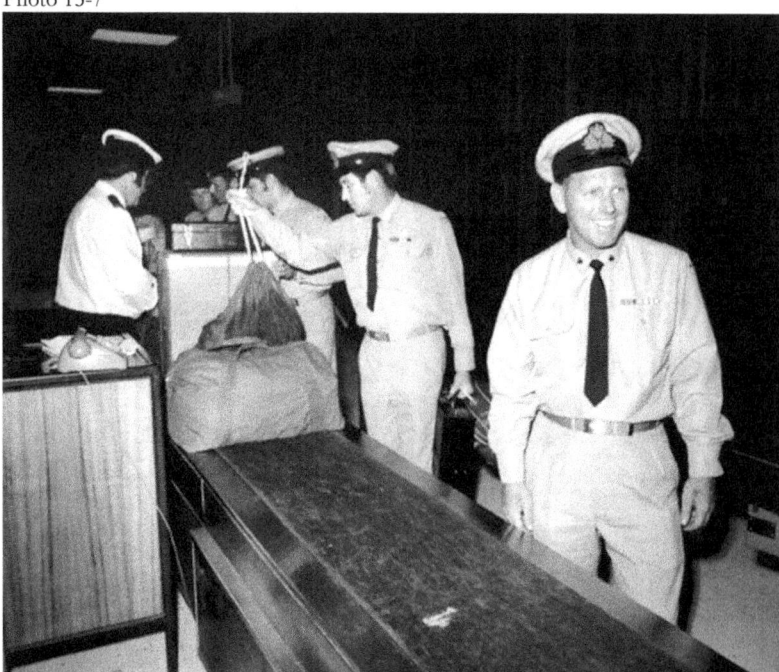

Members of the 8th Contingent, RAN Clearance Diving Team 3 arriving at Mascot Airport, Sydney, after returning from their tour of duty in Vietnam. Lt. Edward ("Jake") Wilfred Linton is followed by CPOCD John Joseph Gilchrist, POCD Phillip Charles Narramore, ABCD Anthony Leonard Ey, and ABCD Brian John Furner, 5 May 1971. Jake Linton collection

16

Hospital Ships *Repose* and *Sanctuary*

The outstanding record of mission accomplishment compiled by the USS Sanctuary *in vital service to the Nation has earned her the highest praise. The magnificent professionalism of her officers and men was responsible for the saving of human life, both military and civilian.*

As senior commander of the United States land forces in this northernmost region of the Republic of Vietnam, I consider it my duty and an honor to express a debt of gratitude on behalf of all those combat soldiers whom the USS Sanctuary *has so gallantly served. Her record deserves no less honor than the award recommended.*

—Endorsement by Lt. Gen. James W. Sutherland Jr., USA, regarding award of the Meritorious Unit Commendation to the hospital ship USS *Sanctuary* (AH-17).[1]

Photo 16-1

USS *Sanctuary* at Mare Island Naval Shipyard in 1967.
USS *Sanctuary* (AH-17) 1967 cruise book

Chapter 16

Two hospital ships served in Vietnam, USS *Repose* (AH-16) and USS *Sanctuary* (AH-17). Constructed as merchant ships, the Navy ordered them converted to hospital ships near the end of World War II.

Repose was launched as SS *Marine Beaver* at the Sun Shipbuilding and Dry Dock Company, Chester, Pennsylvania, on 8 August 1944, and delivered to the Navy prior to completion for conversion to a hospital ship. She was commissioned USS *Repose* on 26 May 1945. Ordered to the Far East, she exited the Panama Canal on 14 July to become an active unit of the U.S. Pacific Fleet. She served with Service Squadron Ten at Shanghai, China, from 30 September 1945 to 15 October 1946, with the exception of one week spent at Tsingtao, China.[2]

Photo 16-2

USS *Sanctuary* (AH-17) anchored at Wakayama Harbor, Japan, on 15 September 1945, embarking patients who had been prisoners of war. Absent in the photo are later improvements, including a helicopter platform aft, installation of TACAN (tactical air navigation used by military aircraft), and a new paint scheme.
USS *Sanctuary* (AH-17) 1967 cruise book

Following a repair and modernization period at San Francisco, *Repose* returned to Tsingtao, joining sister ship USS *Benevolence* (AH-13). Only a two-week availability at Yokosuka for repairs, and nine days at Shanghai, interrupted two years of continuous service from 1 March 1947 until 29 April 1949, at the North China Station. In April 1949, *Repose* received casualties from the heavy cruiser HMS *London*, and the frigates HMS *Black Swan* and HMS *Amethyst*—victims of attack by the Communist Chinese.[3]

HMS *AMETHYST* INCIDENT ON THE YANGTZE RIVER

> *The casualties were:* H.M.S. London, *13 killed, 15 wounded;* H.M.S. Consort, *10 killed, 4 seriously wounded,* H.M.S. Amethyst, *19 killed, 27 wounded;* H.M.S. Black Swan, *7 wounded. In addition, 12 ratings are still missing. Of the damage to the ships, the* London *suffered the most severely, having been holed repeatedly in her hull and upper works. The damage to the* Consort *and the* Black Swan *was less serious.* London *and the* Black Swan *have already completed their emergency repairs. The* Amethyst *suffered severe damage but was repaired by the efforts of her own crew to be capable of a speed of seventeen knots.*

—First Lord of the Admiralty George Henry Hall, speaking to the House of Lords on 26 April 1949, regarding the circumstances in which His Majesty's ships were fired upon in the Yangtze River.[4]

Photo 16-3

HMS *Amethyst* leaving Chinnampo on the west coast of North Korea, following enemy bombardment of the town's port installations, date unknown.
Australian War Memorial photograph 042345

On the morning of 20 April, while proceeding from Shanghai up the Yangtze River to Nanking, and some sixty miles short of her destination, *Amethyst* had come under heavy fire from Chinese Communist Force batteries hidden in reed beds on the north bank. She suffered considerable damage and casualties from the unprovoked attack, and eventually grounded on Rose Island (Leigong Dao). Her commanding officer, Lt. Comdr. Bernard Moreland Skinner, RN, landed about sixty of his crew, including the wounded; who got ashore by swimming or in sampans, being shelled and machine-gunned as they did so. A large proportion, with Chinese help, were able to reach Shanghai.[5]

The destroyer HMS *Consort* was ordered from Nanking to assist *Amethyst*; and *Black Swan* from Shanghai to Kiang Yin, forty miles downriver from her position. *Consort* arrived in mid-afternoon but, heavily hit, was unable to take *Amethyst* in tow, and continued downriver. *London* was ordered to proceed up the Yangtze and meet *Black Swan* and *Consort* at Kiang Yin around 2000. *Consort* being too damaged to participate further, was ordered to Shanghai, to land her dead and wounded and effect repairs.[6]

Amethyst was refloated by her own efforts at 0200 on the 21st, despite her hull being holed in several places, and anchored two miles above Rose Island. There were only four unwounded officers on board; her captain having been severely wounded, her first lieutenant wounded, and her doctor killed. Later that morning, *London* and *Black Swan* tried to close her, but came under heavy fire (which was returned) and suffered some casualties. Both ships returned to Kiang Yin where they were fired upon again. Damaged and with additional casualties, they proceeded to Shanghai.[7]

That evening, a RN officer and RAF doctor reached *Amethyst* by Sunderland flying boat. Additional wounded were evacuated that night, and the frigate moved ten miles upriver to evacuate more. She was left with three naval officers, one RAF doctor, 52 ratings and 8 Chinese on board. On the night of the 22nd, Lt. Comdr. John Simon Kerans, RN, Assistant Naval Attaché at Nankin, arrived to assume command. *Amethyst* moved a further four miles upriver, where she remained for three months before escaping on the night of 30 July. The destroyer HMS *Concord* was present when this occurred.[8]

CONSIGNED TO THE RESERVE FLEET

Repose was operated by the Military Sea Transportation Service with a civilian crew, from 3 September 1949 to 26 August 1950, when she resumed her Navy role. A new Navy crew sailed her from Yokosuka to Pusan, Korea, where she embarked 189 United Nations casualties for return to Japan. *Repose* was returned to commissioned status on 28 October 1950. Following Korean War service, the hospital ship was decommissioned at Hunters Point Naval Shipyard, San Francisco, on 21 December 1954. She was consigned to the Reserve Fleet at nearby Suisun Bay for the next ten and one-half years, until once again called to active duty.[9]

VIETNAM WAR SERVICE

Photo 16-4

Hospital ship USS *Repose* (AH-16) anchored in Da Nang Harbor.
USS *Repose* (AH-16) Republic of Vietnam 1969-1970 cruise book

Repose returned to commissioned status on 16 October 1965. She departed San Francisco on 3 January 1966 and, after refresher training and upkeep in Pearl Harbor and Subic Bay, arrived at Chu Lai on 14 February. Vietnam marked the innovation of the use of mobile hospital support offshore. *Repose* positioned herself near the sites of the heaviest battles, and took aboard virtually all casualties via helicopter. She was responsible for the I Corps Tactical Zone (stretching north from Da Nang to the DMZ), was normally under way, and seldom left the combat zone.[10]

In earlier conflicts, hospital ships often received wounded from greater distances, by slower overland or water transport, resulting in delayed treatment and increased loss of life. When possible, the ships entered the nearest safe port to embark casualties. This occurred on 29

April 1949, when *Repose* sailed with 77 British casualties and 118 American evacuees from Shanghai for Hong Kong. In appreciation, HMS *London* presented her with a plaque. Earlier, in his remarks to the House of Lords, George Hall had acknowledged U.S. naval assistance provided, but did not mention the hospital ship—likely because word of her involvement had not yet reached the Admiralty.

> I should mention that the United States naval authorities at Shanghai placed their resources unstintingly at our disposal, and the kindness and help of the British communities at Shanghai have been beyond all praise.[11]

30 January 1967 marked *Repose*'s 3,000th consecutive helicopter landing, and 29 March, her 2,000th surgical operation in Vietnam. USS *Sanctuary* arrived in Vietnam in April to share the ever-increasing workload of combat casualties.[12]

Photo 16-5

Rapid movement of casualties airlifted to *Repose* by helicopter for triage and treatment. USS *Repose* (AH-16) Republic of Vietnam 1969-1970 cruise book

HOSPITAL SHIP *SANCTUARY*

> *Then, from the* NEW JERSEY *came the blast of the bugle. Each man of the mighty battleship snapped to the right hand salute. The men of* SANCTUARY,

> *including all [former] prisoners [of war] able to stand, returned the honors. Then across the water from the* NEW JERSEY *came the roar of three rousing cheers. Then, down the line, as* SANCTUARY *steamed slowly by one after another of the escort ships, each one followed the example of the flagship, manned their rails and cheered.*
>
> —Actions initiated by the battleship *New Jersey*, flagship of Adm. Raymond A. Spruance, USN, commander Fifth Fleet, as *Sanctuary* stood out of the harbor at Wakayama, Japan, in September 1945, carrying former Allied prisoners of war to Naha, Okinawa. After receiving the customary request sent by a Navy ship upon meeting another commanded by a more senior officer, "Permission to proceed on duty assigned," and answering in reply, "Affirm," the *New Jersey* paid homage to the brave men aboard *Sanctuary*, who had suffered mighty at the hands of their Japanese captors.[13]

Sanctuary, a sister ship of *Repose*, had also been launched at the Sun Shipbuilding and Dry Dock Company, Chester, Pennsylvania. She was laid down as SS *Marine Owl*, but the Navy, anticipating large numbers of casualties from the planned forthcoming invasion of Japan, ordered her converted into a hospital ship. Operation OLYMPIC—code word for the invasion of Kyushu, the southernmost main Japanese island—was scheduled for November 1945. The operation did not take place because Japan surrendered in August, immediately following the atomic bombing of Hiroshima and Nagasaki.[14]

Sanctuary was fitted out at the Todd Shipbuilding Company yard, Hoboken, New Jersey, and commissioned on 20 June 1945. She was at Pearl Harbor two months later, but hostilities had ended during her transit there. She was ordered to Wakayama, Japan, to embark liberated prisoners of war. By 15 September, 1,139 Americans, British, and Australians had been squeezed aboard—most captured at the fall of Singapore, Hong Kong, or in the Battles of Java.[15]

Following a stormy passage to Naha, Japan, with 40 knot winds and heavy seas, *Sanctuary* delivered her patients to the Army; then sailed for Nagasaki, on the northwest coast of Kyushu. There, her crew witnessed firsthand the devastation of the second bomb dropped on Japan. The hospital ship also journeyed to Buckner Bay, Okinawa, and Apra Harbor, Guam, bringing off at each, repatriated prisoners of war. Finally, *Sanctuary* sailed for the United States and was decommissioned at the Philadelphia Naval Shipyard on 15 August 1946.[16]

VIETNAM WAR SERVICE

On 1 March 1966, *Sanctuary* was taken from the Reserve Fleet at Philadelphia (where she had lain idle for nearly twenty years) and towed to the Norfolk Naval Shipyard, Portsmouth, Virginia, for reactivation. In June, she was moved a second time to Avondale Shipyard on the Mississippi River, north of New Orleans, for extensive modernization. Simultaneously, a majority of ship's company and her hospital company underwent pre-commissioning training at Naval Base, Norfolk, and the Portsmouth Naval Hospital.[17]

Photo 16-6

Vice Adm. Robert B. Brown, MC, USN, the Surgeon General of the U.S. Navy, inspecting members of the hospital company during pre-commissioning training at U.S. Naval Hospital, Portsmouth, Virginia.
USS *Sanctuary* (AH-17) 1967 cruise book

Sanctuary was recommissioned on 15 November 1966 at New Orleans, Capt. John F. Collingwood in command. Hospital ships are unique in the U.S. Navy in that they have two commanding officers. The commanding officer of the USS *Sanctuary* was Captain

Collingwood, and of Naval Hospital, USS *Sanctuary*, Capt. Gerald J. Duffner, MC, USN. (Photographs of their counterparts aboard *Repose* follow on the next page, as well as three members of *Repose*'s nursing staff attired in uniforms of that era.)[18]

Photo 16-7

At left, Capt. John F. Collingwood, USN, commanding officer, USS *Sanctuary*; on the right, Capt. Gerald J. Duffner, MC, USN, commanding officer, naval hospital. USS *Sanctuary* (AH-17) 1967 cruise book

Following modernization, *Sanctuary* boasted a heliport, three x-ray units, a blood bank, an artificial kidney machine, ultrasonic diagnostic equipment, a recompression chamber, and other modern equipment to support her twenty wards and four operating rooms. These many improvements enabled her mission to shift in emphasis. It changed from that of an "ambulance" ship carrying wounded and sick to hospitals in rear areas, to that of a fully equipped hospital carrying medical facilities close to the combat area. Her ship's complement was 70 officers and 498 enlisted, with another 316 medical personnel assigned to staff the embarked naval hospital.[19]

Photo 16-8

At left: Capt. Robert Frederick Menge, USN, commanding officer, USS *Repose*; right: Capt. Arthur J. Draper, MC, USN, commanding officer, naval hospital aboard ship. USS *Repose* (AH-16) Republic of Vietnam 1969-1970 cruise book

Photo 16-9

LCDR Mary C. Lukacs   LTJG Carol A. Crandal   LTJG Janice M. Quinn
Surgery Department Nursing staff aboard USS *Repose*.
USS *Repose* (AH-16) Republic of Vietnam 1969-1970 cruise book

Sanctuary arrived off Da Nang on 10 April 1967, and quickly began treating wounded casualties. On her first day in the combat zone, she received ten burn cases resulting from the explosion of a land mine under an amphibious tractor. Through 29 April, she provided medical treatment along the South Vietnamese coast, including the areas of Chu Lai, Phu Bai, Dong Ha, and Da Nang. On that day, she received orders to proceed to an area off Tam Ky to provide medical support for Operation BEAVER CAGE.[20]

Sanctuary arrived on station on 1 May and within two days, received 130 casualties, of which approximately 110 were Marines wounded in action. Two-and-a-half weeks later, she was just off the coast of the Demilitarized Zone (DMZ) on 18 May to support Operation BEAU CHARGER. These operations were but two of numerous amphibious operations and raids carried out by the Seventh Fleet's Amphibious Ready Group and Marine Corps Special Landing Force in Vietnam. The below map from the cruise book of the amphibious assault ship USS *Okinawa* (LPH-3), shows the proximity of these landings to the DMZ.[21]

Map 16-1

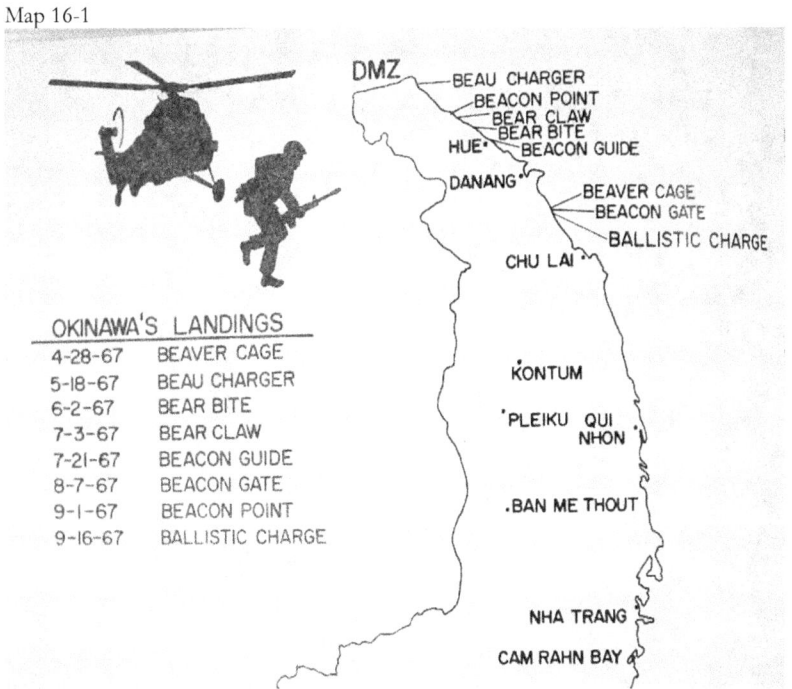

Amphibious landings in which the *Okinawa* took part in 1967
USS *Okinawa* (LPH-3) Western Pacific 1967 cruise book

On 18 May (her thirty-ninth day "on the line"), *Sanctuary* accepted her 1,000th patient. On 21 May, she reached her highest daily patient census: a staggering 634 patients on board. Following her relief on station by *Repose* on 23 May, *Sanctuary* proceeded to Da Nang, to provide additional medical support for the area. By the end of her first period on the line, in fifty-four days the medical ship had:

- Admitted 1,368 patients
- Discharged 698 to duty
- Transferred 117 by medical air evacuation

During this same period, there were 625 helicopter landings aboard *Sanctuary* to receive and discharge patients; 800 surgical operations; 1,950 units of whole blood administered; and 8,387 X-rays taken.[22]

Photo 16-10

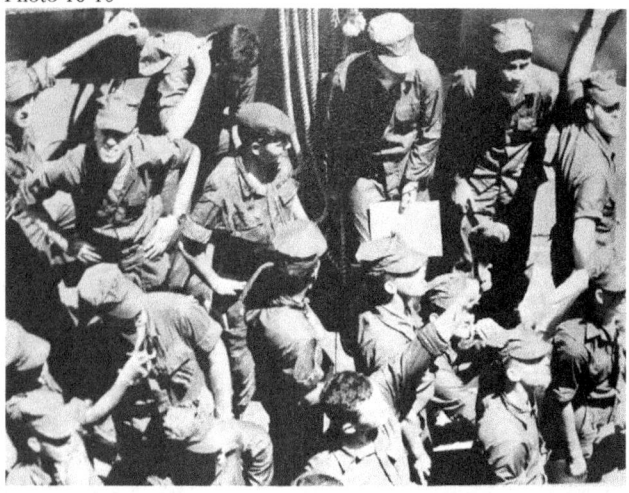

Patients recovered sufficiently to return to their units, being taken ashore by landing craft. The more seriously wounded are being sent ashore for air evacuation back to the United States for further treatment.
USS *Sanctuary* (AH-17) 1967 cruise book

On 3 June, *Sanctuary* departed the combat zone, bound for Subic Bay with 553 patients aboard, arriving two days later for a 10-day upkeep period. By 17 June, she was again off South Vietnam. On 2 July, the hospital ship was engaged in treating casualties from Operation CIMARRON, involving fierce fighting southwest of Con Thien, a U.S. Marine Corps combat base located near the DMZ. By 10 July, the patient census on board had reached 614.[23]

Photo 16-11

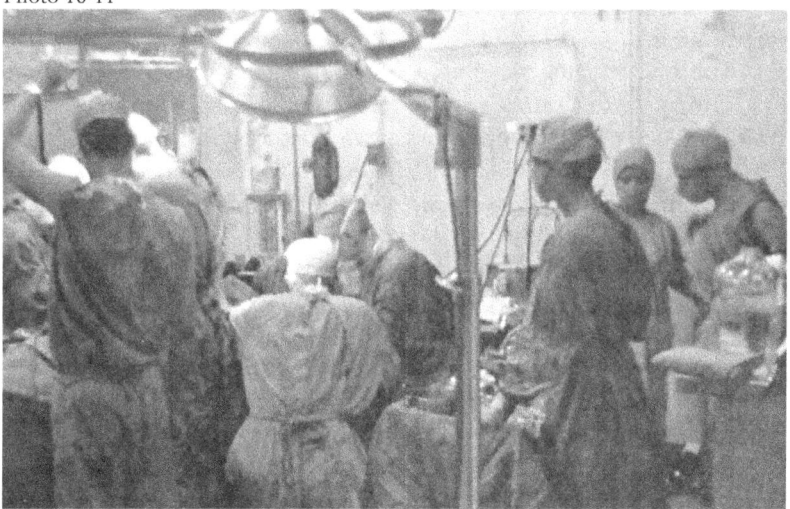

Operating room aboard USS *Repose*.
USS *Repose* (AH-16) Republic of Vietnam 1969-1970 cruise book

Later that month, *Sanctuary*'s crew and medical staff received a welcome repose, a visit to Hong Kong for some rest and relaxation. The ship moored to Alfa No. 1 buoy on 19 July for a five-day stay. Liberty ashore was unrestricted on Hong Kong Island and the Kowloon Peninsula, except for commander Seventh Fleet curfew, which required all personnel to be off the street between 0100-0500. As the ship had neared Hong Kong, many patients made "miraculous" recoveries, and over 200 of them were also allowed to enjoy liberty ashore.[24]

By the morning of 26 July, *Sanctuary* was back on the line at Da Nang, and once again fell into her now familiar routine as the command came over the 1 M.A. announcing system, "Naval Hospital, Man Your Patient Handling Stations, Section 2 Provide."[25]

HOSPITAL SHIPS PUT OUT OF SERVICE

Repose served twenty-three periods on the line between 16 February 1966 and 31 January 1970, and was awarded Navy Unit Commendations for the periods 22 February 1966 to 8 February 1967, and 9 February 1967 to 1 April 1969. *Sanctuary* garnered a Navy Unit Commendation for the period 10 April 1967 to 10 April 1969, and a Meritorious Unit Commendation for 11 April 1969 to 14 April 1971. Her Vietnam service stretched from 9 April 1966 through 2 May 1971, encompassing twenty tours on the line. (A copy of *Sanctuary*'s MUC may be found at Appendix E.)

The two hospital ships continued their Vietnam service into the early 1970s. After treating more than 9,000 casualties and admitting over 24,000 patients for inpatient care in Southeast Asian waters, *Repose* departed Vietnam on 14 March 1970, bound for the United States. She was decommissioned in May, and laid up in the Pacific Reserve Fleet, San Francisco Group.[26]

Sanctuary (the only Navy hospital ship off Vietnam after 16 March 1970) increased her busy schedule through 1970 and into 1971. On 23 April 1971, she departed Da Nang for the United States, with visits to Hong Kong, Sasebo, and Pearl Harbor en route to San Francisco. She was decommissioned on 15 December 1971 at the Hunters Point Naval Shipyard—but would serve again.[27]

She was recommissioned on 18 November 1972, minus the red cross on her white hull. Her new mission would be to serve primarily as an overseas dependent support ship, providing medical care, plus limited commissary and exchange facilities to authorized personnel and their dependents.[28]

Photo 16-12

The crew of USS *Sanctuary* board the ship during her recommissioning ceremony on 18 November 1972. She was the first United States Navy ship to have an integrated crew, comprised of male and female officers and enlisted personnel.
USS *Sanctuary* (AH-17) 15 December 1971-14 December 1973 cruise book

17

Mobile Shipyards

Photo 17-1

Repair ship USS *Jason* (AR-8) lying at anchor off Vung Tau, in 1968, with the coastal minesweeper USS *Woodpecker* (MSC-209) moored starboard aft, while refueling.
Naval History and Heritage Command photograph #NH 107752

The Seventh Fleet's repair ships (AR) and destroyer tenders (AD) served as mobile repair facilities, with highly skilled personnel on board to perform any repair that might arise. Without their service in the Western Pacific, many shipyards and ship repair facilities would have been overloaded with work. Consequently, important repairs needed by ships serving in the Tonkin Gulf and off Vietnam would have been delayed until time permitted their completion.[1]

The repair ships and tenders provided repair support in deep-water ports near the combat zone, principally at Subic Bay, Philippines; Sasebo and Yokosuka, Japan; Kaohsiung, Taiwan; and Vietnam itself. As shown in the table, of the nine ARs and ADs, the repair ships *Ajax*, *Hector*, and *Jason* spent the most time in Vietnamese waters. Between 8 June 1968 and 16 July 1972, each was sent to Vung Tau (and *Hector* also Da Nang) three to five times. The remaining repair ships and tenders were dispatched to Vietnam a single time. However, their service to the fleet at other locations provided ships rotating "off the line" repair support, concurrent with rest and relaxation outside the combat zone.[2]

All of the ARs and ADs, with the exception of *Samuel Gompers*, were of World War II vintage. Commissioned in July 1967, the bright and shiny 645-foot *Gompers* boasted an 85-foot beam, resulting in her crew affectionately referring to her as "Fat Sam."

Repair Ships and Destroyer Tenders that Served in Vietnam

Ship	Comm/Decom	Periods in the Combat Zone
Ajax (AR-6) *Vulcan*-class	3 Oct 43 31 Dec 86	8-23 Jun 1968; 26 Sep-11 Oct 1969; 18 Apr-11 May 1970; 25 Aug-23 Sep 1971; 29 Oct-21 Nov 1971
Hector (AR-7) *Vulcan*-class	7 Feb 44 31 Mar 87	24 Jul-18 Aug 1970; 21 Feb-16 Mar 1972; 14-30 Apr 1972; 12-16 Jul 1972
Jason (AR-8) *Vulcan*-class	19 Jun 44 24 Jun 95	25 Jul-16 Aug 1968; 27 Dec 1969-21 Jan 1970; 23 Mar-13 Apr 1971
Delta (AR-9) *Delta*-class	16 Jun 41 20 Jun 70	29 Jun-27 Jul 1969
Klondike (AR-22) *Klondike*-class	30 Jul 45 15 Sep 74	31 Mar-24 Apr 1969
Markab (AR-23) *Hamul*-class	15 Jun 41 19 Dec 69	3-22 Nov 1967
Piedmont (AD-17) *Dixie*-class	5 Jan 44 30 Sep 82	29 Jun-8 Jul 1972
Isle Royale (AD-29) *Shenandoah*-class	9 Jun 62 11 Mar 71	9 Jan-1 Feb 1967
Samuel Gompers (AD-37) *Samuel Gompers*-class	15 Jul 67 27 Oct 95	21-30 Apr 1972

ISLE ROYALE FIRST TENDER SENT TO VIETNAM

The first "mobile shipyard" (AR or AD) sent to Vietnam was the *Isle Royale*. She had been laid down on 16 December 1944, and launched 19 September 1945, but then, with the war over, laid up in the Pacific Reserve Fleet, Long Beach, California, until commissioned on 9 June 1962. The 492-foot destroyer tender arrived at Chu Lai in early 1967 to retrieve salvageable equipment and machinery from the tank landing ship *Mahnomen County* (LST-912). *Isle Royale* entered Vietnamese waters on 9 January, and departed on 1 February. The intervening three weeks were devoted to the rigorous work of stripping ship.³

While anchored off Chu Lai on 1 January, *Mahnomen County* had dragged anchor and turned beam to the beach. Heeled over by heavy waves, she broached, grounding on rocks inside the surf zone. Seventh Fleet salvage forces (assisted by Harbor Clearance Unit One, Team 1, from Da Nang) had carried out intensive, but unsuccessful efforts to save the stricken ship. All efforts to free *Mahnomen County* were thwarted by a rock ledge extending seaward along her full length. Moreover, salvage efforts were hindered by northeast monsoon winds as high as 40 knots, and an 18-foot plunging surf.⁴

Photo 17-2

Chu Lai, South Vietnam, August 1967.
Naval History and Heritage Command photograph #NH 74346

As a result of the impact caused by the grounding and heavy surf working her against the rocks, *Mahnomen County* suffered extensive damage. All spaces below her third deck were holed and open to the sea, and the tank deck was completely cracked athwartships at Frame 22. The main deck was cracked in several areas, and the bow doors were sprung open and their foundations cracked. Salvage efforts were terminated on 31 January. *Mahnomen County* was decommissioned and, stripped of usable equipment, her hull was demolished by personnel of the Naval Support Activity Da Nang, detachment at Chu Lai.⁵

MARKAB VISITS VIETNAM IN NOVEMBER 1967

Later that year, the repair ship *Markab* arrived at Vung Tau to provide services to ships and craft there. She had been laid down at Ingalls Shipbuilding, Pascagoula, Mississippi, as SS *Mormacpenn* for the Moore-McCormack Line. She was completed in 1941, acquired by the U.S. Navy on 2 June 1941, and subsequently converted and commissioned the attack cargo ship USS *Markab* (AKA-31). She was redesignated a destroyer tender (AD-21) on 27 September 1942, and saw much duty in the Pacific Theater. Mothballed and returned to service twice after the war, *Markab* was recommissioned a repair ship (AR-23) in 1960.

Photo 17-3

Repair ship USS *Markab* under way, location and date unknown.
USS *Markab* (AR-23) Western Pacific 1967 cruise book

Markab left Alameda, California, on 7 June 1967 on deployment. She provided services at Subic Bay, Kaohsiung, Yokosuka, Vung Tau, and Subic again, before returning to homeport on 16 December 1967.

Map 17-1

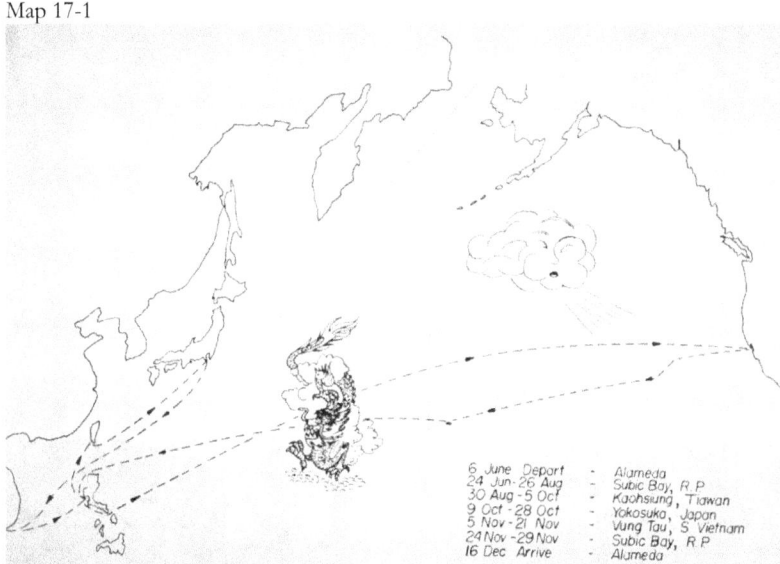

Track chart of repair ship USS *Markab*'s 1967 Western Pacific deployment
USS *Markab* (AR-23) Western Pacific 1967 cruise book

Markab's Repair Department encompassed nearly three dozen shops, spread among five divisions. The titles of these shops evidence the diversity of service she provided fleet units.[6]

R-1 Division: Hull Repair
- Shipfitter Shop
- Sheet Metal Shop
- Weld and Blacksmith Shops
- Pipe Shop
- Canvas Shop
- Pattern Shop
- Carpenter Shop

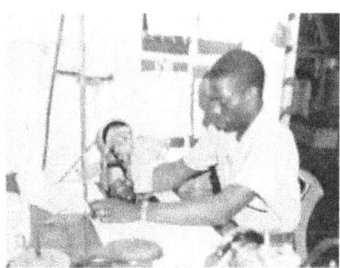

R-2 Division: Machine Repair
- Light Machine Shop
- Heavy Machine Shop
- Outside Repair and Refrigeration Shops
- Valve and Pump Shops
- Internal Combustion Engine Shop
- Foundry
- Central Tool Room
- Grind Shop
- Boiler Repair Shop

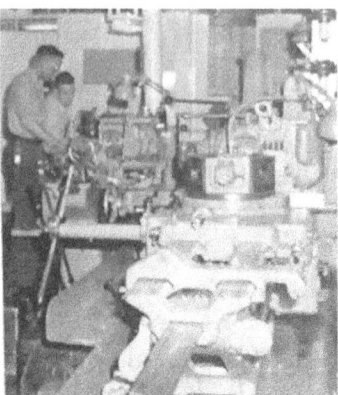

R-3 Division: Electrical Repair
- Electrical Shop
- Gyro Shop
- Projector Shop
- Battery Shop
- Meter Repair Shop

R-4 Division: Electronics Repair
- Electronics Repair Shop
- Fleet Calibrations Lab
- Teletype Repair Shop
- Crypto Repair Shop

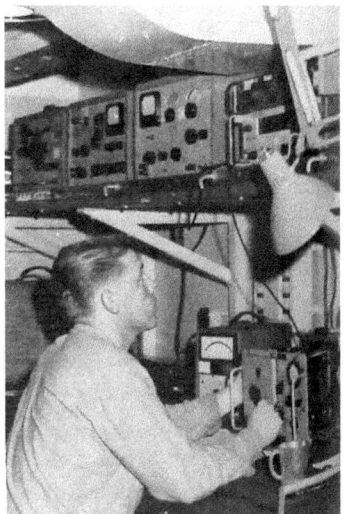

R-5 Division: Ordnance Repair
- Optical Shop
- Print Shop
- Watch Repair Shop
- Gauge Shop
- Typewriter Shop
- Draft Shop
- Diving Locker
- Ordnance
- Fire Control

Amid the busy period at Vung Tau, *Markab* sailors enjoyed some liberty in the coastal city.

Photo 17-4

USS *Markab* sailors inspecting a local delicacy at a vendor's stand in Vung Tau.
USS *Markab* (AR-23) Western Pacific 1967 cruise book

AJAX PROVIDES SERVICE AT VUNG TAU IN 1968

Photo 17-5

Popular beach at Vung Tau, South Vietnam.
USS *Klondike* (AR-22) Western Pacific 1969 cruise book

Ajax arrived at Vung Tau on 9 June 1968. Although the seaport was a rest and recreation center for allied forces, her crew worked without break for thirteen days straight, making badly needed repairs and providing services to ships and small craft, as well as to various Army and Air Force equipment ashore. Combat damage to PBRs and MSBs (river patrol boats and minesweeping boats) was fairly common, as were requirements to mend and return them to duty.[7]

Vung Tau offered beaches, and other leisure and nightlife activities for soldiers, sailors, and Marines enjoying interludes from combat duty, but lay adjacent to the Viet Cong-infested Rung Sat Special Zone. Known from before the French colonization of Vietnam as a refuge for the pirates and bandits that preyed upon river and coastal traffic, the area later became one of the strongholds of the Binh Xuyen gangsters (sometimes referred to as the "Vietnamese Mafia"). American servicemen commonly referred to the Rung Sat as the Forest of Assassins, because the Vietnamese word *rung* can be translated as "forest, jungle, or woods," and *sat* as "assassin, killer, or murderer."[8]

Merchant vessels making passage to Saigon with cargoes vital to the allied war effort, had to traverse the Rung Sat. The Viet Cong, determined to interdict shipping, planted remotely-detonated mines on the Saigon River bottom. After observing the movements of deep draft vessels, the enemy laid mines in the areas of the river through which ships had to traverse to navigate safely. Upon sighting a ship entering a "kill zone," an observer hidden in heavy foliage along the bank would detonate the mine, via an electrical signal transmitted through a cable to the river-emplaced weapon.

Map 17-2

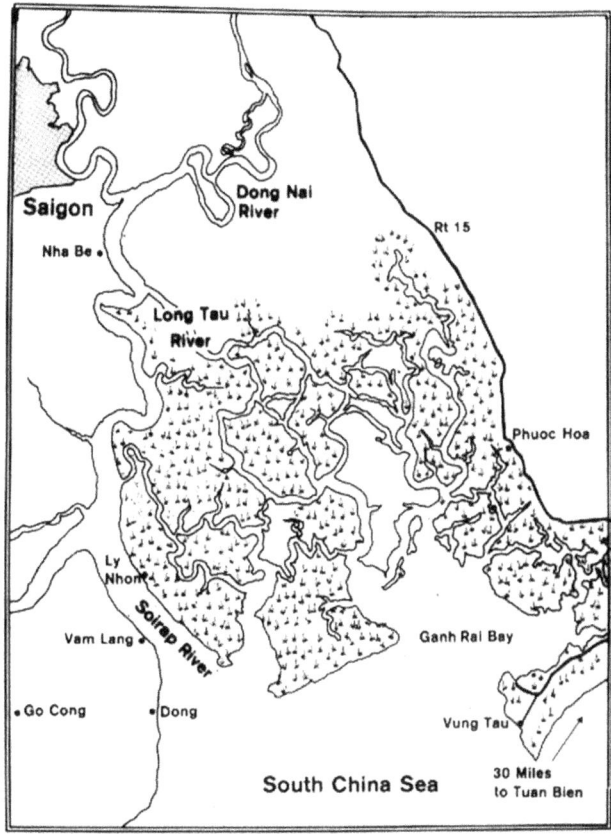

Rung Sat Special Zone (Soirap is more commonly spelled Soi Rap)
Naval History and Heritage Command photograph #NH 96343

To counter the threat the mines presented, the Navy deployed a detachment of 57-foot wooden minesweeping boats in 1966 to Nha Be, sited at the junction of the Soi Rap and Long Tau rivers, the main waterways between the port of Saigon and the South China Sea. Located just seven miles south of Saigon, Nha Be would become a major combat and logistics hub during the course of the war.[9]

In *A Soldier Reports*, General William C. Westmoreland explains his rationale for ordering minesweeping of the waterway between the South China Sea and Saigon, as well as the Army's efforts to break the Viet Cong's grip on the surrounding area:

I was long concerned that the VC might sink a large vessel along the forty-mile course of the Saigon River to block the vital shipping channel between Saigon and the sea. The main channel ran through fifty square miles of mangrove swamps and thousands of tributary waterways, a region known as the Rung Sat, which was an ideal base from which the VC could operate against shipping. One of the most savage pieces of terrain in the world, the Rung Sat has almost no ground that is not subject to inundation from a six-foot tidal variation. The houses of the few villages in the region are built on stilts.

To break the VC hold on the region, I later shifted a battalion of the 1st Infantry Division to the Rung Sat where the men encountered some of the most unusual and trying conditions ever faced by American soldiers. Because the men spent hours and even days patrolling in water up to the waist, companies had to be rotated frequently to forestall foot and skin diseases. The men slept at night on air mattresses and at particularly high tides might awake to find themselves afloat. Wooden platforms built above the high-tide marks served as helicopter pads. It was a strange war within a strange war, but it cut one of the main VC infiltration routes into the Saigon region and disrupted the enemy's organization. Even though the VC later hit an occasional ship with rocket, machine gun, or mortar fire, they never succeeded in blocking the shipping channel to Saigon.[10]

Daily sweeping for mines was both tedious and dangerous, particularly in the meandering, restricted passages that cut through the Rung Sat Swamp. Here, boat crews, as well as Navy and merchant ships in passage, regularly faced death from command-detonated mines, or from ambush by Viet Cong armed with rockets, 57mm and 75mm recoilless rifles, and automatic weapons.[11]

SCRAMBLE THE "SEAWOLVES"

In addition to the minesweeping boats and river patrol boats, assault craft of the Mobile Riverine Force carrying soldiers on search and destroy missions also operated in the Rung Sat, and elsewhere in the Mekong Delta. When under attack, they could count on armed Navy helicopters—the "Seawolves" of HAL-3 (Helicopter Attack, Light, Squadron Three)—to rapidly respond to urgent requests for support. Mine Division 112 at Nha Be was supported by HAL-3 Detachment 2 aircraft sporting rocket pods and machine guns.[12]

Photo 17-6

A UH-1 Iroquois helicopter ("Seawolf") from Light Attack Squadron Three (HAL-3) flies over an abandoned Viet Cong village on the Cua Lon River, July 1969. National Archives photograph #USN 1140788

CONTINUED SERVICE OF *AJAX*

Ajax departed Vung Tau for Subic Bay on 22 June 1968. After arriving there three days later, she undertook a repair job of considerable importance: the regunning of four 5-inch mounts on the heavy cruiser *Boston* (CAG-1). Work around the clock for seven days returned *Boston* to ready status.[13]

Over the next three years, *Ajax* returned to Vung Tau four times to perform additional repair work: a two-week stay from 27 September to 10 October 1969; servicing Vung Tau from 13 April to 9 May 1970 in support of the American offensive in Cambodia; and two stints there in 1971, in September and for the first three weeks in November.[14]

JASON FOLLOWS *AJAX* AT VUNG TAU

> *I was standing midwatch. I was up on the bridge of the ship. It was raining and storming outside. I was pretty much up there by myself, just the officer that was on duty and myself. I was probably feeling a little on the lonely side. When you're watching a mid-watch like that, all you do, at least in this particular situation, you had to look at a radar screen to look for any approaching aircraft or other ships or whatever. But, that's in the middle of the night and there's not that much going on. So, I had my pen and my paper there and I was just fooling around with different ideas. I had the title in my mind for quite a long time. I just liked the sound of the title "Rhythm Of The Rain." And it's rather interesting because that title isn't sung anywhere in the song. I mean, the first line of the song is "listen to the rhythm of the falling rain."*
>
> —John Claude Gummoe, former USS *Jason* crewmember and later lead singer with the 1960s band The Cascades, in an interview with Gary James.[15]

Many interesting and talented people have served, and are serving today in the U.S. Navy—three such were crewmembers of the repair ship *Jason* in the early 1960s. Among the ship's complement of 53 officers and 1,244 enlisted men were John Claude Gummoe, David Wilson, and Leonard Green. Many ships had their own bands, formed by members of their crews, as did the *Jason*. These three individuals, and others who later joined them, formed The Silver Strands (a reference to Silver Strand Beach, a sand-spit that forms the outer edge of the San Diego Bay), later to become The Thundernotes, and finally, The Cascades.[16]

The origin of The Cascades began with Dave Wilson and John Gummoe sitting around the fantail of *Jason* singing harmony to the guitar playing of Len Green, mostly singing the songs of the Everly Brothers, The Lettermen, and The Four Freshman. This was before "the British Invasion"—a cultural phenomenon of the mid-1960s, when rock and pop music from the United Kingdom became popular in the United States. The three shipmates began performing in Navy Service Clubs at bases in San Diego. Wilson was a drummer and singer, and Green played guitar and sang, as well.[17]

A talented composer, Green helped inspire "Rhythm of the Rain," written by John Gummoe aboard the *Jason* while she was in transit from San Diego to Japan, and released in November 1962. Gummoe was the lead singer and played the keyboards. At the close of 1963, Billboard Magazine ranked "Rhythm of the Rain" the third largest selling single in the world. In 1999, Broadcast Music, Inc. listed it as the ninth most played song on American radio and television in the 20th century. Having grown tired of living out of a suitcase, Gummoe left the band in 1967 to pursue a solo career.[18]

Crewmembers, like bandmembers, come and go, as Navy ships continue operating. *Jason*'s deployments, following the departure of the original members of The Cascades (at the completion of their enlistments or duty aboard her), included three periods at Vung Tau. The following dates reflect when *Jason* entered and departed the combat zone in carrying out these assignments:

- 25 July - 16 August 1968
- 27 December 1969 - 21 January 1970
- 23 March - 13 April 1971

1969 VISITS BY *KLONDIKE* AND *DELTA*

Klondike provided services at Vung Tau during the first three weeks in April of her 1969 Western Pacific deployment. Her crew of 25 officers and 600 men principally performed repairs to the hull, machinery, electrical, electronic, and ordnance equipment of other ships, but could also complete battle and operational damage repairs to ships and craft. Other services included medical and dental, fleet logistics support, and command facilities for a Service Squadron commander and staff.[19]

The repair ship *Delta* was at Vung Tau in July 1969, lying at anchor in the harbor while repairing "shot up" river boats.[20]

HECTOR'S SERVICE AT VUNG TAU IN 1970 AND 1972

Photo 17-7

Repair ship USS *Hector* (AR-7) at Vung Tau in 1972.
Courtesy of NavSource (photograph from the collection of Sy Blau)

Repair ship *Hector* left Long Beach on 23 June 1970 on deployment. She arrived at Sasebo on 10 July. Ten days later, she sailed for Vung Tau, where she provided services to the fleet until her departure on 16 August—for which she received her first Vietnam Service Medal. Two years later, while deployed and serving as flagship for commander, Service Group Three, she spent three periods in the combat zone: 21 February-16 March, 14-30 April, and 12-16 July 1972. During these periods, she provided services at Vung Tau and Da Nang.[21]

"FAT SAM" AND *PIEDMONT'S* VUNG TAU SERVICE

Samuel Gompers arrived at Da Nang on 9 April 1972, and left there on the 16th for Subic Bay. She returned to Da Nang a second time on 22 April for another short stint, before proceeding back to the Philippines at month's end. *Piedmont* (AD-17) was at Da Nang a little over a week, two months later. She entered South Vietnamese waters on 29 June, and left on 8 July 1972, the last of the three destroyer tenders to depart Vietnam. *Hector*, the last repair ship, left Vietnam on 16 July 1972.[22]

Photo 17-8

At left: Caricature of "Fat Sam" from *Samuel Gompers'* 1971-72 cruise book. Top right: USS *Samuel Gompers* (AD-37) providing tender services to four destroyers at Subic Bay, May 1969. Bottom right: USS *Piedmont* (AD-17) at sea, circa the 1970s.
Navy History and Heritage Command photographs #USN 1139041 and #NH 107745

LAURELS FOR THREE REPAIR SHIPS

Ajax was awarded a Meritorious Unit Commendation for the period 1 July 1968 to 1 October 1969. Her award citation reads in part "in providing vital repairs and logistic services to ships and craft of the United States and other free-world nations engaged in combat operations in Southeast Asia." Both she and *Jason* received a Navy Unit Commendation for their service with Naval Support Activity, Saigon; for the periods 2-29 October 1969 and 29 December 1969 to 10 January 1970, respectively. (A copy of the NUC award citation may be found in Appendix F.) *Hector* also received a MUC, which reads in part, "For meritorious service from 26 January to 25 August 1972 in direct support of United States SEVENTH Fleet combat operations in Southeast Asia. USS *HECTOR* contributed materially to the success of these operations by rendering vital fleet repair services to United States and friendly naval forces operating in the Republic of Vietnam."[23]

18

Ammunition Ships

Praise the Lord and Pass the Ammunition.

—Title of an American patriotic song published as sheet music in 1942 by Famous Music Corp. Frank Loesser wrote the song in response to the Japanese attack on Pearl Harbor on 7 December 1941.

Photo 18-1

Ammunition ship USS *Mount Hood* (AE-11) explodes in Seeadler Harbor, Manus, Admiralty Islands, 10 November 1944.
U.S. Naval History and Heritage Command photograph #NH 96173

Twenty-four ammunition ships served in Vietnam. The bulk of them—five *Lassen*-class, seven *Mount Hood*-class, and two ex-*Andromeda*-class attack cargo ships—were of World War II vintage. Comprising the remaining ten were two *Suribachi*-class AEs, commissioned in 1956 and 1957; three *Nitro*-class AEs, commissioned in 1959; and five AEs of the *Kilauea*-class, commissioned between 1968 and 1971.

OLDEST AMMUNITION SHIPS

Photo 18-2

Ammunition ship USS *Mount Baker* (AE-4) at sea, circa 1961.
Naval History and Heritage Command photograph #2015

Mount Baker, *Rainier*, and *Shasta*, the three oldest AEs sent to Vietnam, were commissioned in 1941. The Navy's pre-WWII war plans had called for wartime requirements for auxiliaries to be met by the conversion of merchant ships rather than by new construction. Accordingly, as the world crisis deepened, the sea service acquired seven diesel-powered C2 cargo ships, under construction in builders' yards, or soon to be, for conversion to ammunition ships. The motor vessel MV *Shooting Star*, became the USS *Lassen* (AE-3), lead ship in the *Lassen*-class of ammunition ships.[1]

The 459-foot former freighters, displacing 14,225 tons, could make 15 knots, propelled by two Nordberg diesel TSM219 engines coupled to a single Falk main reduction gear and single propeller shaft. Ship's complement was 21 officers and 260 enlisted. The five ships that served in Vietnam are identified in the table, as well as the span of their service in Vietnam; when they were first commissioned, and decommissioned for the last time; and unit awards received.

Ship Name/ Span of Vietnam Service	Comm/ Decom	Unit Commendations
Mauna Loa (AE-8) 29 Nov 67-11 Apr 68 (6 tours)	27 Oct 43 26 Feb 71	
Mazama (AE-9) 12 May 66-23 Oct 69 (13 tours)	10 Mar 44 May 70	
Mount Baker (AE-4)/ex-*Kilauea* 13 Feb 66-12 Jun 69 (24 tours)	16 May 41 2 Dec 69	
Rainier (AE-5) 4 Jul-24 Oct 65 (7 tours)	21 Dec 41 7 Aug 70	
Shasta (AE-6) 4 Nov 66-24 Apr 67 (8 tours)	9 Jul 41 1 Jul 69[2]	MUC: 3 Nov 66-16 Apr 67

Mount Baker was commissioned USS *Kilauea* (AE-4). Her name was later changed to *Mount Baker* on 17 March 1943 to avoid confusion with a similarly named ship, the unclassified vessel USS *Kailua* (IX-71), employed as a cable laying ship. All five of the *Lassen*-class AEs were decommissioned after World War II, or the Korean War, or both, as part of post-war downsizing. They were all returned to service before Vietnam (as shown below). *Rainier* was the first to serve in the war zone, with her initial "swing on the line" beginning on 4 July 1965. The final tour of *Mazama*, the last to serve, ended on 23 October 1969.

Ship	Recommissioned	Ship	Recommissioned
Mount Baker	5 December 1951	*Mauna Loa*	27 November 1961
Rainier	25 May 1951	*Mazama*	27 November 1961
Shasta	15 July 1953		

FIRST EAST COAST AE DISPATCHED TO VIETNAM

Mazama departed the United States on 17 March 1966 to relieve the *Wrangell* (AE-12) in Vietnam. Based at Davisville, Rhode Island, about eighteen miles south of Providence, she was the first East Coast ammunition ship deployed to the war zone. When *Mazama* arrived in Subic Bay on 5 May, her crew (16 officers and 187 enlisted) was still inexperienced in their ship's primary mission of transferring ammunition at sea. Practice came early, as five days later, *Mazama* was en route to her operating area off the coast of Vietnam.[3]

During her first tour on the line, *Mazama* averaged 68.5 short tons of ammunition per hour in CVA (attack carrier) rearming. After much practice in the Tonkin Gulf, her crew increased the transfer rate to 138 and finally to 146 short tons during the last two efforts on station. In total, she rearmed carriers 41 times, including those listed below, as well as 74 other ships, of which 49 were destroyers.

- USS *Enterprise* (CVAN-65)
- USS *Kitty Hawk* (CVA-63)
- USS *Handcock* (CVA-19)
- USS *Ranger* (CVA-61)
- USS *Constellation* (CVA-64)
- USS *Oriskany* (CVA-34)
- USS *F. D. Roosevelt* (CVA-42)
- USS *Coral Sea* (CVA-43)[4]

Photo 18-3

Ammunition ship USS *Mazama* alongside a CVA (attack aircraft carrier).
USS *Mazama* (AE-9) Western Pacific 1966 cruise book

Mazama returned to Subic Bay for the final time on 29 October. She departed on 3 November, empty of ammunition, to begin her journey home the long way—west via the Suez Canal. For the initial part of the voyage, she was not alone. *Shasta*, also an East Coast ship, had arrived at Subic just days earlier. The two proceeded in company for a while, which offered time for training. *Shasta* rehearsed her approaches and Burton riggings, made some token transfers, said farewell, and changed her course to Vietnam. *Mazama* headed for the equator where, in a traditional crossing-the-line ceremony, 180 pollywogs would be inducted by King Neptune into the "Solemn Mysteries of the Ancient Order of the Deep."[5]

On 20 December 1966, *Mazama* arrived back at Davisville, following a nine-month, around-the-world deployment. She was welcomed by fireboats as she entered the Narragansett Bay, with relatives and friends waiting on the pier to see their loved ones home for the Christmas holidays.[6]

Photo 18-4

Sailors at Davisville, Rhode Island, welcoming USS *Mazama* home.
USS *Mazama* (AE-9) Western Pacific 1966 cruise book

FORMER ATTACK CARGO SHIPS (AKA)

Photo 18-5

USS *Virgo* (AE-30) under way in Subic Bay, Philippines, 26 August 1970.
Naval History and Heritage Command photograph #NH 97300

Ex-*Andromeda*-class Attack Cargo Ships
459 feet; 14,200 tons; 16.5 kts; complement: 52 officers, 421 men
two boilers, one De Laval steam turbine, single propeller (6,000shp)

Ship Name/ Span of Vietnam Service	Comm/ Decom	Unit Commendations
Chara (AE-31)/ex-AKA-58 22 Nov 66-16 Nov 71 (29 tours)	14 Jun 44 10 Mar 72	NUC: 24 Apr-27 Nov 71
Virgo (AE-30)/ex-AKA-20 12 Feb 66-4 Nov 70 (28 tours)	10 Jul 43 18 Feb 71	

Two of the ammunition ships that served in Vietnam were former attack cargo ships, both products of Federal Ship Building and Dry Dock Company, Kearney, New Jersey. *Virgo*, the oldest, was commissioned on 16 July 1943. During the latter part of World War II, she earned seven battle stars in the Pacific Theater for participation in island assault operations and campaigns, including those at Tarawa, Kwajalein, Hollandia, Guam, Peleliu, Iwo Jima, and Okinawa.[7]

During her war service, Thomas Heegan served as communications officer aboard the *Virgo*, and later transmogrified her into the fictious USS Reluctant, in his 1946 novel *Mr. Roberts*. In an ensuing screenplay, he wrote that the cargo ship sailed "from Apathy to Tedium with occasional side trips to Monotony and Ennui." The 1955 comedy-drama film *Mister Roberts*, starring Henry Fonda, James Cagney, William Powell, and Jack Lemmon (based on the novel) was nominated for three Academy Awards, with Lemmon winning the award for Best Supporting Actor.[8]

Virgo earned another nine battle stars in the Korean War. Later, as part of post-war downsizing, she was laid up in the Columbia River Group, Pacific Reserve Fleet, Astoria, Oregon, on 3 April 1958, and struck from the Naval Register on 1 July 1961. As the numbers of carriers, cruisers, destroyers, and amphibious ships in the Tonkin Gulf and off South Vietnam increased, so too did the requirement for Service Force support. Accordingly, *Virgo* was withdrawn from the National Defense Reserve Fleet in September 1965, reinstated in the Naval Register on 1 November 1965, and recommissioned an ammunition ship, USS *Virgo* (AE-30), on 19 August 1966.[9]

Sister ship *Chara* was commissioned on 14 June 1944. She earned four battle stars in World War II—at Lingayen Gulf, Leyte, and Manila Bay in the Philippines, and later at Okinawa—and another seven in Korea. *Chara* was reconfigured as an ammunition ship during the early part of the Korean War, but was not reclassified as such until reactivated in 1965 for service in Vietnam.[10]

Chara completed twenty-nine swings between 22 November 1966 and 16 November 1971, and was awarded a Meritorious Unit Commendation by Adm. Elmo R. Zumwalt Jr., chief of Naval Operations, for her last deployment. Zumwalt, former commander, Naval Forces Vietnam, had become, at age forty-nine, the youngest officer ever to serve as the Navy's top officer. (A copy of *Chara*'s award citation may be found in Appendix G.)[11]

Photo 18-6

Adm. Elmo R. Zumwalt Jr., chief of Naval Operations (left), and Rear Adm. Robert S. Salzer, commander, Naval Forces Vietnam, chat aboard an aircraft following a visit to Nam Can Naval Base, South Vietnam. Salzer had relieved Vice Adm. Jerome H. King Jr. (Zumwalt's successor as ComNavForV) a month earlier, in April 1971.
Naval History and Heritage Command photograph USN 1148801

MOUNT HOOD-CLASS AMMUNITION SHIPS

All seven 459-foot *Mount Hood*-class ships served in Vietnam. They were originally of the *Wrangell*-class, which was renamed following the obliteration of USS *Mount Hood* (AE-11) and everyone aboard her by a massive explosion. (Depicted in the photograph on the opening page of this chapter.) On 10 November 1944, *Mount Hood* was lying at anchor off Manus in the Admiralty Islands, north of New Guinea, loaded with ammunition. Her crew was busy that morning moving ammunition in all five of her holds. Eighteen men had left the ship at 0830 to make a run ashore. Twenty minutes later, the remaining 249 men aboard *Mount Hood* disappeared in a towering, mushroom-shaped cloud of smoke, from which fragments of steel and shrapnel began falling on and around nearby ships and craft.[12]

Comdr. Chester Gile, USNR (Retired), who witnessed the massive devastation wrought by the explosion, later described the incident in an account published in the U.S. Naval Institute *Proceedings*, February 1963:

> *Mount Hood* was anchored in approximately 35 feet of water. The force of the explosion blasted a trench in the harbor bottom, reported by divers as 1000 feet long, 200 feet wide and 85 feet maximum depth. In the trench was found the largest piece of the ship's hull - a piece less than 100 feet in its longest dimension. Destruction was complete. Nothing was found after the explosion except fragments of metal which struck other ships. There were no bits of human remains, no supplies of any kind, nothing that had been made of wood or paper, with the single exception of a few tattered pieces of a signal notebook, floating on the water several hundred yards away.[13]

The horrific loss of life, served as a reminder then, and today, of the potential danger associated with duty aboard an ammunition ship.

Mount Hood-class (formerly *Wrangell*-class)
459 feet; 14,960 tons; 14 kts; complement: 267
three boilers, one GE turbine, single propeller (6,000shp)

Ship Name/ Span of Vietnam Service	Comm/ Decom	Unit Commendations
Diamond Head (AE-19) 5 May 67-28 Mar 68 (8 tours)	9 Aug 45 1 Mar 73	MUC: 16-23 Oct 70
Firedrake (AE-14) 24 Nov 65-6 Sep 70 (19 tours)	27 Dec 44 Oct 70	
Great Sitkin (AE-17) 1 May-4 Oct 68 (4 tours)	11 Aug 45 2 Jul 73	
Mount Katmai (AE-16) 6 Jul 65-27 Jan 73 (54 tours)	21 Jul 45 14 Aug 73	MUC: 16 Nov 69-19 Jun 70 MUC: 24 Apr 72-23 Feb 73
Paricutin (AE-18) 4 Jul 65-16 Oct 70 (33 tours)	3 Mar 45 23 Apr 71	
Vesuvius (AE-15) 2 Aug 65-29 Jan 73 (55 tours)	16 Jan 45 14 Aug 73	MUC: 4 Oct 69-7 May 70 NUC: 28 Feb 72-19 Feb 73
Wrangell (AE-12) 17 Nov 65-9 Apr 69 (16 tours)	10 Oct 44 21 Dec 70[14]	

FIRST AEs TO EARN VIETNAM SERVICE MEDALS

One *Lassen*-class AE, five *Mount Hood*-class, and two of the newer *Nitro*-class ships, were the first ammunition ships to earn Vietnam Service Medals, serving "on the line" in 1965. All were active when the United

States formally entered the Vietnam War; *Great Sitkin* and *Mount Katmai* having remained active after World War II, and the others having been brought "out of mothballs" in the early 1950s.

Ship	Recommissioned	Ship	Recommissioned
Paricutin	24 June 1950	*Vesuvius*	15 November 1951
Diamond Head	9 August 1951	*Great Sitkin*	Remained active
Firedrake	11 October 1951	*Mount Katmai*	Remained active
Wrangell	14 November 1951		

Photo 18-7

Loads of 750-pound bombs and 2.75-inch rockets being highlined from USS *Paricutin* (AE-18) to USS *America* (CVA-66), during operations in the Tonkin Gulf, 9 June 1968. National Archives photograph #USN 1135492

Many ships served in Vietnam before eligibility for Vietnam Service Medals began, earning Armed Forces Expeditionary medals instead. One such was *Rainier*. While deployed to the Western Pacific in 1964, Subic Bay was initially the focal point of her 7th Fleet support activities. When the Tonkin Gulf crisis occurred, 4-5 August, *Rainier* immediately put to sea and proceeded to the gulf to rearm carriers conducting strikes on North Vietnamese bases. Through late October, she operated between Subic Bay and replenishment areas off Vietnam. Near deployment's end, she sailed for Japan and in December 1964, arrived back at her homeport, Concord, California.[15]

In the late spring of 1965, *Rainier* resumed operations in the Tonkin Gulf, along with the *Paricutin* and *Nitro*. By January 1966, she had transferred at sea almost 12,000 tons of ammunition, 83 tons of freight, and 11,500 pounds of mail. In February, she returned to Concord.[16]

MOUNT HOOD-CLASS SHIPS' UNIT AWARDS

Three *Mount Hood* AEs earned four Meritorious Unit Commendations and one Navy Unit Commendation between them, during their service in Vietnam.

- *Vesuvius* (AE-15)   MUC: 4 October 1969-7 May 1970
- *Vesuvius* (AE-15)   NUC: 28 February 1972-19 February 1973
- *Diamond Head* (AE-19)   MUC: 16-23 October 1970
- *Mount Katmai* (AE-16)   MUC: 16 November 1969-19 June 1970
                           MUC: 24 April 1972-23 February 1973

(Mount *Katmai*'s MUC citation may be found at Appendix H.)

Photo 18-8

USS *Vesuvius* leaving the San Francisco Golden Gate Bridge astern.
USS *Vesuvius* (AE-15) Western Pacific 1969-1970 cruise book

Vesuvius' 1969-1970 deployment illustrates the level of effort that led to her first unit award. On 17 September 1969, after completing her pre-deployment loadout at the Concord Naval Weapons Station, she sailed on a cold and foggy day on the nineteenth deployment of her then-25-year career. *Vesuvius* stopped at Pearl Harbor briefly, then continued onward for the Far East. After evading a typhoon en route, and delivering some ammunition to Yokosuka, Japan, she reached Subic Bay on 17 October, a full month after departing the California coast.[17]

Preparations for duty in Vietnam began immediately. Over the next six-and-a-half months, *Vesuvius* would make eight stops at the busy Subic Naval Operating Base for the purpose of effecting repairs and taking aboard ammunition, food, and spare parts. Ninety-eight days were devoted to seven demanding line runs in the combat area off Vietnam. Working day and night, often under adverse weather conditions, the AE's hardworking crew safely handled 12,505 tons of ammunition while rearming 108 ships alongside. All commitments were met, and extraordinarily fine performance was the norm.[18]

SURIBACHI- AND *NITRO-*CLASS SHIPS

The *Suribachi-* and *Nitro-*class AEs are often considered to be of the same class, because of their common builder and similar characteristics. All five ships were products of Bethlehem Steel Shipyard, Inc, Sparrows Point, Baltimore, Maryland, in the mid- to late 1950s. Several years after entering service, *Suribachi* and sister ship *Mauna Kea* were fitted with FAST (Fast Automatic Shuttle Transfer System), which enabled them to transfer a "bird" from their hold to the magazine of a missile-firing ship in 90 seconds. They also each received a helicopter platform on their fantail, enabling vertical replenishment of fleet units.[19]

Suribachi-class Ammunition Ships
512 feet; 15,688 tons; 20 kts; complement: 20 officers, 324 enlisted
two boilers, steam turbines, single shaft (16,000shp)

Ship Name/ Span of Vietnam Service	Comm/ Decom	Unit Commendations
Mauna Kea (AE-22)	30 Mar 57	MUC: 2-Mar-6 Oct 68
24 Jan 67-25 Jan 73 (43 tours)	30 Jun 95	MUC: 10 Jun 70-29 Jun 71
Suribachi (AE-21)	17 Nov 56	MUC: 5 Jul 72-27 Feb 73
12 Jun 72-8 Feb 73 (12 tours)	2 Dec 94	

Photo 18-9

USS *Suribachi* (AE-21) under way in 1957.
Naval History and Heritage Command photograph #NH 97245

Nitro (AE-23) was the lead ship in the first new class of AEs to be built from the keel up as ammunition ships. She could make 19 knots, and keep far ranging fleets supplied with a wider variety and a larger quantity of ammunition. Three of her five cargo holds were outfitted for stowing palletized ammunition loads, and each had two cargo elevators for rapid break out of cargo, which provided redundancy in case of an elevator failure. The use of battery-powered fork trucks for moving palletized loads, in conjunction with the elevators, provided safer, more efficient handling of ammunition.[20]

Nitro-class Ammunition Ships
512 feet; 15,688 tons; 19 kts; complement: 28 officers, 331 enlisted
Two boilers, one steam turbine, single propeller (16,000shp)

Ship Name/ Span of Vietnam Service	Comm/ Decom	Unit Commendations
Haleakala (AE-25) 20 Dec 65-13 Aug 72 (51 tours)	3 Nov 59 10 Dec 93	MUC: 10 Nov 71-25 Aug 72
Nitro (AE-23) 30 May 72-29 Jan 73 (12 tours)	1 May 59 28 Apr 95	MUC: 18 May 72-17 Feb 73
Pyro (AE-24) 4 Jul 65-31 Oct 72 (47 tours)	24 Jul 59 31 May 94	MUC: 6 Nov 69-15 May 70 NUC: 5 Dec 64-23 Oct 65 NUC: 21 Jan-12 Nov 72

Photo 18-10

Battleship USS *New Jersey* (BB-62) receiving cans of powder for her 16-inch guns from the USS *Haleakala* (AE-25), 4 October 1968. Plywood sheets protect her wooden decks.
National Archives photograph #USN 1141175

NEW *KILAUEA* SHIPS JOIN THE FLEET IN EARLY 70s

Photo 18-11

USS *Kilauea* (AE-26) off Quonset Point, Rhode Island, 25 September 1969.
National Archives photograph #USN 1141257

USS *Kilauea* (AE-26), the first of a new class of AEs, was commissioned at Boston Naval Shipyard, Charleston, Massachusetts, on 10 August 1968. Her commanding officer, Capt. William L. McGonagle, USN, had been awarded the Medal of Honor, for heroism while in command of USS *Liberty* (AGTR-5) when Israeli aircraft and three motor torpedo boats attacked the technical research ship on 8 June 1967.[21]

Photo 18-12

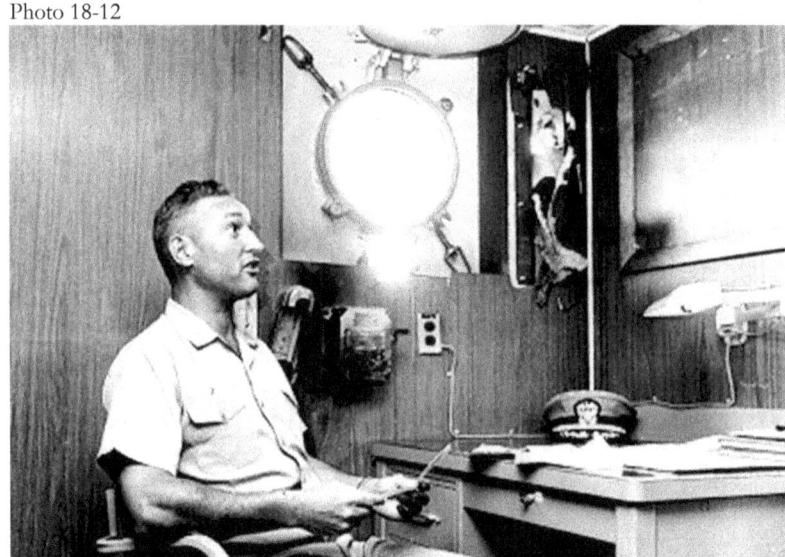

Comdr. William L. McGonagle, commanding officer, USS *Liberty* (AGTR-5), in his cabin aboard ship, 11 June 1967. The holes in the bulkhead were part of the damage suffered when Israeli forces attacked the *Liberty* off the Sinai Peninsula on 8 June. Naval History and Heritage Command photograph NH 97474

Kilauea-class Ammunition Ships
564 feet; 19,940 tons; 20 kts; complement: 28 officers, 375 enlisted
three boilers one GE Electric turbine, single propeller (22,000shp)

Ship Name/ Span of Vietnam Service	Comm/ Decom	Unit Commendations
Butte (AE-27)	14 Dec 68	
6 Feb-3 Mar 73 (3 tours)	3 Jun 96	
Flint (AE-32)	20 Nov 71	
4 Nov 72-22 Mar 73 (4 tours)	4 Aug 95	
Kilauea (AE-26)	10 Aug 68	MUC: 20 Aug 71-29 Jun 72
7 Aug 70-12 Jun 72 (20 tours)	1 Oct 80	
Mount Hood (AE 29)	1 May 71	MUC: 7 May 72-30 Mar 73
13 May 72-13 Mar 73 (12 tours)	10 Aug 99	
Santa Barbara (AE-28)	11 Jul 70	MUC: 29 Jun 72-31 Jan 73
5 Jul 72-18 Jan 73 (9 tours)	25 Sep 98	

Five of the new ammunition ships served in Vietnam. *Kilauea*, lead ship of her class, began her first swing on the line on 7 August 1970. Sister ship *Flint* was the last AE to serve on the line, with her final tour ending on 22 March 1973. President Richard Nixon announced on 15 January 1973, the end of offensive operations against North Vietnam. The Paris Peace Accords ending the conflict were signed twelve days later, followed by the phased withdrawal of the remaining American troops over the next two months.

Ammunition ships continued to support naval combatant forces off the Vietnam coast during the clearance of American mines from Haiphong Harbor in North Vietnam (one of the provisions of the Peace Accords) and difficulties with Hanoi, regarding the return of American prisoners of war. The nine AEs that served off Vietnam in 1973, and their respective final day on the line, are identified below.

Ship	End Date	Ship	End Date	Ship	End Date
Santa Barbara	18 Jan 73	*Nitro*	29 Jan 73	*Butte*	3 Mar 73
Mauna Kea	25 Jan 73	*Vesuvius*	29 Jan 73	*Mount Hood*	13 Mar 73
Mount Katmai	27 Jan 73	*Suribachi*	8 Feb 73	*Flint*	22 Mar 73

Capt. Philip M. Palmer, USN (Retired) authored the closing pages of this chapter which, written in the first person, provide a very interesting and compelling account of the many interrelated actions and hard work that were necessary to keep the ships on the gunline supplied with ammunition, and naval aircraft with bombs and mines. During his tour as commanding officer, U.S. Naval Magazine, Subic Bay, Palmer witnessed first-hand, three phases of the war; the buildup, the naval bombardment and mining campaigns, and the drawdown.

MINING OF HAIPHONG HARBOR AND LINEBACKER OPERATIONS AGAINST NORTH VIETNAM

On the morning of 8 May 1972, three Marine A-6 Intruders and six Navy A-7 Corsair attack planes from the carrier *Coral Sea* (CVA-43) flew toward the coast of North Vietnam. Shortly afterward, the aircraft laid strings of thirty-six 1,000-pound Mark 52 mines in the water approaches to Haiphong, through which most of North Vietnam's imported war material and all of its fuel supply passed. During succeeding months, other carrier aircraft dropped thousands of mines and 500-pound, Mark 36 Destructors in the seaways of North Vietnam's secondary ports and remined the Haiphong approaches.[22]

The mining campaign immobilized twenty-seven Sino-Soviet bloc merchant ships in Haiphong not eager to risk a transit of the mined waters. It, in conjunction with U.S. air attacks on North Vietnam's supply lines ashore, helped cut short the enemy's "Easter Offensive," an invasion of South Vietnam launched by the North Vietnamese Army on 30 March 1972. Eventually, the mining operation and the U.S. counter-offensive "Linebacker" bombing campaign induced the North Vietnamese to negotiate an end to the war.[23]

NAVAL MAGAZINE, SUBIC BAY

> *The primary munitions of the DDs, 5" projectiles, were being produced in the U.S., airlifted to the red label area of Cubi Point NAS, transported by truck to the pier for loading aboard a waiting AE [ammunition ship], and thence to the combat zone. We estimated that less than two weeks after manufacture, a 5" projectile was out the bore of a DD 5" gun mount and on its way into Vietnam.*
>
> —Capt. Philip M. Palmer, USN (Retired), former commanding officer, Naval Magazine, Subic Bay, remarking on the high consumption of gun rounds by destroyers on the gunline.

Photo 18-13

Naval Magazine, Subic Bay (on Camayan Point), Philippine Islands, September 1965. Naval History and Heritage Command photograph #NH 74182

LINEBACKER OPERATIONS SUPPORT

Capt. Philip M. Palmer, USN (Retired) shares in the closing pages of the chapter, a first-hand account of the efforts of Naval Magazine, Subic Bay to support U.S. Naval combat operations in Vietnam:

When I arrived in July 1971, the tempo of operations in South Vietnam was building and so were the naval force levels in Southeast Asia. To interdict coastal supply trains into the South from North Vietnam, the Navy established a coordinated coastal shore bombardment from destroyers, cruisers, the battleship *New Jersey*, and carrier-based aircraft. All of the ordnance to support these operations flowed through NAVMAG Subic. The firing rates were such that the supply train could barely keep up with the expenditures. More logistic support force ships were ordered to the area, many deployed from the East Coast.

We were short of manpower and ordnance handling equipment (called "yellow gear" because of its color) and so an augmentation was begun. Our SeaBee sailors (naval construction battalion personnel) who operated and maintained the yellow gear were increased in number, and every piece of yellow gear not required either ashore or afloat elsewhere in the Navy was relocated to Subic. Our gear all had to be certified for handling explosive material so we got the very best. We also received additional sailors and U. S. civilians and hired additional Filipino nationals.

By early 1972 we were operating at a full wartime tempo. We had installed an additional pontoon pier at Camayan Wharf for use by ordnance-transporting LSTs (tank landing ships). We routinely were servicing five ships simultaneously: two at the wharf, one at the pontoon pier, and two at anchor. This went on around the clock even though handling ammunition after dark is not supposed to be done. Subic Bay looked like a scene in WWII movies. Ships were at piers and at anchor throughout the Bay. We had 1,000,000 pounds net equivalent high explosive weight on Camayan Wharf at all times. How do I know that? Because that's all the explosive handling regulations permitted. So, that's what we said we had.

Ordnance demand was so heavy that the ammunition ships (AEs) were being used just as shuttles to carry ammunition to the multi-commodity ships (AOEs) which stayed on station in the Gulf of Tonkin servicing the combatants around the clock.

In July of 1972, we had a series of typhoons pass over Luzon and stack up to the east of the island. The result of this meteorological phenomenon was torrential tropical rain. It rained without letup for 31 days and nights. When it finally ended, 103" of rain had fallen. I remember the number because, coincidentally, during that period, the Magazine moved a record 103,000 tons of ordnance under nearly impossible conditions.

The Central Luzon plain was flooded to the second story level in many places and you could go from the foothills of the mountains below Baguio to Manila by Banca boat. You could not get from Subic to Manila by vehicle. The main road from Cubi down to Subic washed away carrying the main waterline with it and left us in all that rain with no water. The sailors in the Magazine lived down the hill in Subic so they had to come to work via a circuitous road that remained intact. We got water in buckets from tanker trucks we called "water buffalo" that were parked on Captain's Circle, senior officer housing.

I was proud of the officers and sailors and also of the Filipinos who had no national stake in this war but were willing to work hard for their pesos. A stevedore earned $1.15/day. 15 cents of that went back to the stevedore company for a lunch of fish heads and rice. They were an interesting group. Every morning a large contingent of day laborers would assemble at the union hall in Olongapo. LUSTEVCO, the stevedore contractor, would announce the numbers required for the day. That number would be trucked down to the Magazine piers.

During my tour we had an excellent explosives safety record. The water tower that stood in the middle of Nabasan Wharf had a message written in bold letters on its sides and visible at some distance. It was the byword of explosives safety, "ORDNANCE SAFETY PRECAUTIONS ARE WRITTEN IN BLOOD."

We exercised safe operations as rigorously as we could while operating at a wartime tempo. The one exception I know we made was in staging ammo on Nabasan Wharf for loading aboard waiting and incoming ships. The pier explosive limit was 1,000,000 net equivalent pounds of TNT (NET). That limit would not permit us to keep up with the tempo of operations required to support the peak of the naval war. When asked how much ordnance we had on the wharf I always responded 1,000,000 NET. We probably had more than double that but I never really knew. I couldn't allow myself to worry about it.

MINING OPERATIONS SUPPORT

On Palm Sunday in 1972, I was in the Chapel at the Cubi Point Naval Air Station with my family. Just as the sermon was about to start, a messenger came down the aisle and told me I needed to come to the Magazine immediately. At the gate I was directed by the Duty Officer to proceed directly to the Mine Shop. When I arrived, I found that Master Chief Mineman Wheelock, his minemen, newly reported LDO

(Limited Duty Officer) LCDR Dick Anderson, and the other mine shop CPOs and officers were either there or on their way. We had a TOP SECRET message informing us that the U. S. was about to initiate a naval blockade to interdict all supplies destined for North Vietnam and that a major element of the blockade would be the mining of Haiphong Harbor.

Over the next three weeks, we prepared every mine that went into Vietnam waters; some 8,000 in all including destructor kits for installation aboard aircraft carriers in 500lb bombs to convert them into magnetic influence, bottom mines. Scores of the others were very sophisticated mines which required the minemen to make settings for depth, delays, ship counts, self-neutralization, etc., as called for in the mining plan. It was very complex, exacting work which required meticulous performance and record keeping. This was our proudest moment. We knew we were contributing to an ending of hostilities in a very direct way.

WAR'S END AND RECOGNITION

Photo 18-14

Col. Robinson Risner, USAF (waving), and Capt. James Stockdale, USN, two American prisoners of war released by Hanoi, arriving at Clark Air Force Base, Philippines, in February 1973.
National Archives and Records Administration photograph #USN 1155662

In early 1973, the release of American POWs and an end to U. S. involvement in the war was negotiated. The first stop of the POWs was at Clark Air Force Base in the Philippines. We watched it with pride and tears over TV.

On 10 December 1973, the Naval Magazine officers and men stood at formation while Rear Adm. Doniphan B. Shelton, USN, commander, Naval Base, Subic Bay, presented NAVMAG Subic with the Meritorious Unit Citation for support of the war effort in 1972. The citation, signed by the Secretary of the Navy, read:

> In the face of extremely adverse climatic conditions and unusually demanding operational commitments, U. S. Naval Magazine, Subic Bay carried out its highly important mission of munitions support to the Seventh Fleet with outstanding skill and dedication and provided a significant and vital contribution to successful naval operations in the Southeast Asian theater during this period. By their hard work, perseverance and unfailing devotion to duty throughout, personnel of the U. S. Naval Magazine, Subic Bay reflected great credit upon themselves and upheld the highest traditions of the United States Naval Service.

During the period noted in the award, the Naval Magazine supported record numbers of fleet units and merchant ships, and handled munitions tonnages, which more than doubled any previous comparable peak period. At the height of operations, the Naval Magazine provided ordnance support to six aircraft carriers, fifty destroyers, four cruisers, and 23 logistics replenishment ships. Merchant ships serviced more than tripled during this period, averaging over 80,500 tons per month for six consecutive months. The magazine normally handled less than 20,000 tons per month.

19

Fleet Oilers

MATTAPONI *is no longer a young healthy warrior, but an old matron.... Admittedly, it takes a lot of haze grey "makeup" to keep her appearance up, and much tender loving care to keep her heart pumping, but she seems to thrive on age. It becomes her.*

She is not a glamorous ship, and the men who have served her do not have a glamorous story. But that is always the way with the real heroes. Sweat and heat and fatigue and boredom are as much a threat to the human spirit as bullets, but they do not make headlines. It doesn't make good reading in Des Moines that the glamorous warships couldn't operate without 250 men toiling on a twenty-seven-year-old converted merchant ship.

But that makes the story of MATTAPONI*'s men that much more valuable. They have had less to work with, but they have accomplished so much more. They have left home five times in as many years to fight a war halfway across the world, but they have left an enviable record of service in their wake. They haven't left headlines, but they have left hope. At home, people have trouble pronouncing* MATTAPONI, *but in the South China Sea, she is the "Angel of Market Time."*

> —Tribute to USS *Mattaponi*'s (AO-41) crew from the ship's 1968-1969 cruise book, regarding their efforts during a deployment in which they earned a Meritorious Unit Commendation.[1]

Photo 19-1

USS *Mattaponi* (AO-41) near San Diego, California.
National Archives photograph #K-57335

Photo 19-2

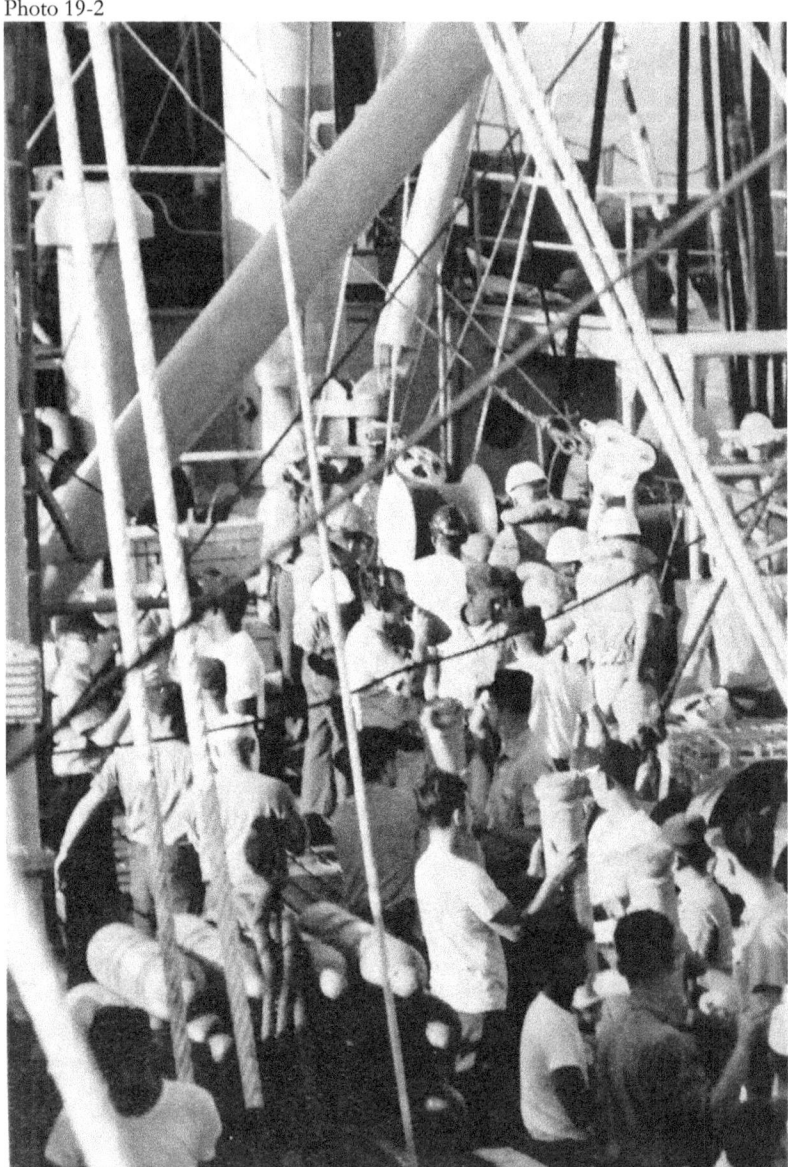

It takes many crewmembers, doing many different jobs, to make an UnRep a success, especially with ships along both sides, and with both highlines in operation.
USS *Mattaponi* (AO-41) Western Pacific 1968-1969 cruise book

Among the Service Force ships that served in Vietnam, fleet oilers (twenty-three in total) were second in number only to the twenty-four ammunition ships. Like the AEs, most of them were also veterans of World War II and the Korean War.

SINGLE *KENNEBEC*-CLASS SHIP IN VIETNAM

USS *Kennebec* (AO-36), the smallest of the fleet oilers, was laid down as SS *Corsicana*, a Maritime Commission type T2-SO tanker, at Bethlehem Steel Shipyard, Sparrows Point, Maryland. The Navy acquired her for conversion on 13 January 1942 (following her launching). She was the single member of the *Kennebec* class to serve in Vietnam, but contributed much to the war effort—completing forty-two "swings on the line" between September 1965 and September 1969.[2]

Photo 19-3

Fleet oiler USS *Kennebec* (AO-36) at sea, location unknown, circa 1964.
National Archives photograph #USN 1106517

Kennebec-class Fleet Oiler
Maritime Commission T2-SO type
501 feet; 22,380 tons; 16.7 kts; complement: 20 officers, 194 enlisted
2 boilers, one Westinghouse turbine, single propeller (12,000shp)

Ship Name/ Span of Vietnam Service	Comm/ Decom	Unit Commendations
Kennebec (AO-36) 25 Sep 65-29 Sep 69 (42 tours)	4 Feb 42 29 Jun 70	

Stretching 502 feet in length, *Kennebec* could carry 134,000 barrels of oil. The twenty-two ships of the other three classes of fleet oilers had varying capacities, as shown in the table, which does not include the *Cimarron*-class ships later "jumboized" to increase their length, displacement, and cargo capacity.

Capacities of the Fleet Oilers that served in Vietnam

Ship Class/ # Ships Vietnam	Year Lead Ship Commissioned	Ship Length	Displ. Tons	Barrels of Oil
Kennebec-class: 1	1942	502 feet	22,380	134,000
Mattaponi-class: 3	1942	520 feet	21,450	135,000
Cimarron-class: 16	1939	553 feet	25,425	146,000
Neosho-class: 3	1954	655 feet	38,000	180,000[3]

MATTAPONI-CLASS FLEET OILERS

Photo 19-4

USS *Tappahannock* (AO-43) transfers aircraft fuel to USS *Enterprise* (CVAN-65), as the cruiser USS *Bainbridge* (CGN-25) comes along the oiler's starboard side to receive mail. Naval History and Heritage Command photograph #L45-274.01.03

The three *Mattaponi*-class AOs—*Mattaponi*, *Neches*, and *Tappahannock*—that served in Vietnam were very similar to *Kennebec*. The three sister ships (the former SS *Kalkay*, SS *Aekay*, and SS *Jorkay*) were all built at Sun Shipbuilding & Drydock Co., Chester, Pennsylvania. Eighteen feet longer that *Kennebec*, they could carry a thousand more barrels of oil than she, but otherwise offered essentially the same capabilities.[4]

Mattaponi-class Fleet Oilers
Maritime Commission T2-A (MC-K) type
520 feet; 21,450 tons; 17.5 kts; complement: 21 officers, 221 enlisted
2 boilers, one Westinghouse turbine, single propeller (12,000shp)

Ship Name/ Span of Vietnam Service	Comm/ Decom	Unit Commendations
Mattaponi (AO-41) 30 Sep 66-27 Jun 70 (25 tours)	11 May 42 30 Sep 70	MUC: 19 Oct 68-15 May 69
Neches (AO-47) 6 Jul 65-30 May 70 (46 tours)	16 Sep 42 1970	
Tappahannock (AO-43) 26 Dec 66-3 Aug 69 (14 tours)	22 Jun 42 6 Mar 70[4]	

NUMEROUS *CIMARRON*-CLASS FLEET OILERS

Photo 19-5

"Jumboized" USS *Ashtabula* (AO-51) under way off San Diego, California.
National Archives photograph #K-84354

Anyone assigned to a U.S. Navy ship during the four decades from the 1940s through the 1970s, likely spent time alongside one or more of the 553-foot *Cimarron*-class oilers. *Cimarron*, at the time of her launching on the morning of 7 January 1939, was the fastest tanker built in the United States and one of the largest in the world.[5]

Of the sixteen *Cimarron*s that served in Vietnam, five of the AOs can be considered members of subclasses. Five converted T3-S2-A3 oilers—*Mispillion, Navasota, Passumpsic, Pawcatuck,* and *Waccamaw*—are often termed the *Mispillion*-class. In addition to being slightly different from other *Cimarron*-class ships constructed with T3-S2-A1 hulls, these ships were "jumboized" from 1964 through 1967 to increase their capacity to 180,000 barrels. This involved cutting the oilers in two, inserting a new mid-body between bow and stern, and fitting a helipad forward on the new, longer ships. All of the "stretched AOs," save *Pawcatuck*, completed tours "on the line" in the South China Sea.[6]

Ashtabula, Caloosahatchee, and *Canisteo* (T3-S2-A3 types as well) were jumboized after the five *Mispillion*s. However, the two groups were quite different in appearance and UnRep equipment. The *Ashtabula*s had several features not shared by the *Mispillion*s. The most prominent of these were a fully enclosed well deck, no exterior deck walkways on the forward superstructure, a tunnel through it to allow the movement of cargo to the forward deck, two sets of STREAM (Standard Tensioned Replenishment Alongside Method) gear, and no helideck on the bow.[7]

Photo 19-6

Bow being joined to USS *Caloosahatchee* (AO-98)
at Bethlehem Steel Company, Baltimore Yards, 16 July 1968.
Naval History and Heritage Command photograph #NH 85038

Of the latter three AOs, only *Ashtabula* served in Vietnam. During her reconfiguration in 1968, a 400-foot midsection was inserted and welded between her bow and stern, replacing the original 310-foot midsection. At completion, she spanned 644 feet in length, displaced 36,500 tons, had a larger liquid cargo capacity of 145,000 barrels, and was manned by a larger crew. Her new complement was 20 officers and 350 enlisted, sailing aboard a ship whose profile now closely resembled that of a modern replenishment oiler (AOR).[8]

Cimarron-class Fleet Oilers
**553 feet; 25,425 tons; 18.3 kts; complement: 34 officers, 267 enlisted
four boilers, two Westinghouse turbines, twin propellers (13,500shp)**

Ship Name/ Span of Vietnam Service	Comm/ Decom	Unit Commendations
Ashtabula (AO-51) jumboized 4 Aug 65-30 Jun 72 (44 tours)	7 Aug 43 30 Sep 82	MUC: 21 May-7 Dec 69 HS: 29-30 Apr 75
Cacapon (AO-52) 18 Mar 66-22 Mar 73 (48 tours)	21 Sep 43 Aug 73	MUC: 8 Oct 69-8 Apr 70
Caliente (AO-53) 4 Jul 65-10 Mar 73 (35 tours)	12 Oct 43 unknown	MUC: 24 Jan-11 Aug 72
Chemung (AO-30) 24 Dec 65-29 Jan 70 (38 tours)	3 Jul 41 18 Sep 70	
Chipola (AO-63) 22 Dec 65-30 Sep 72 (51 tours)	30 Nov 44 14 Aug 73	MUC: 19 May-21 Nov 67 MUC: 30 May-16 Sep 68 MUC: 16 Dec 68 MUC: 17 May-1 Jun 69 MUC: 23 Sep 70-23 Mar 73
Cimarron (AO-22) 22 Jul 65-13 Sep 67 (25 tours)	20 Mar 39 1 Oct 68	
Guadalupe (AO-32) 13 Feb 66-10 Jul 72 (56 tours)	19 Jun 41 1974	MUC: 14 Oct 71-3 Aug 72
Manatee (AO-58) 13 Jul 66-18 Feb 73 (52 tours)	12 Feb 44 14 Aug 73	MUC: 22 Aug 69-9 Feb 70 MUC: 27 May 72-15 Mar 73
Marias (AO-57) 19 Jan-20 Mar 73 (4 tours)	12 Feb 44 2 Oct 73	MUC: 5 May-1 Sep 70
Mispillion (AO-105) jumboized 30 Apr 67-2 Nov 72 (49 tours)	29 Dec 45 26 Jul 74	MUC: 14 Apr-16 Nov 67 MUC: 23 Aug 69-28 Feb 70 MUC: 27 Feb-18 Nov 72
Navasota (AO-106) jumboized 15 Sep 65-2 Jun 72 (58 tours)	27 Feb 46 13 Aug 75	
Passumpsic (AO-107) jumboized 24 Oct 66-12 Feb 73 (55 tours)	1 Apr 46 24 Jul 73	MUC: 27 Nov 67-23 Jun 68 MUC: 5 May 72-27 Feb 73
Platte (AO-24) 12 Jul 65-15 Feb 72 (38 tours)	10 Dec 39 early 1970s	MUC: 4 Nov 69-21 Apr 70
Taluga (AO-62) 4 Jul 65-19 Sep 71 (18 tours)	25 Aug 44 4 May 72	

Tolovana (AO-64)	24 Feb 45	MUC: 29 Sep 72-16 May 73
4 Jul 65-16 Mar 73 (57 tours)	15 Apr 75	
Waccamaw (AO-109) jumboized	25 Jun 46	MUC: 23 Sep-2 Oct 70
8 Jul 72-5 Jan 73 (11 tours)	24 Feb 75	MUC: 27 Jun 72-17 Jan 73

Photo 19-7

USS *Kawishiwi* (AO-146) under way off Oahu, Hawaii.
National Archives photograph #USN 1093776

Neosho-class Fleet Oilers
655 feet; 38,000 tons; 20 kts; complement: 324
two boilers, two turbines, two shafts (28,000shp)

Ship Name/ Span of Vietnam Service	Comm/ Decom	Unit Commendations
Hassayampa (AO-145)	19 Apr 55	MUC: 20-22 Jul 69
10 Jul 65-25 Jan 73 (64 tours)	17 Nov 78	
Kawishiwi (AO-146)	6 Jul 55	MUC: 19 Apr-8 May 75
21 Dec 65-9 Sep 72 (42 tours)	10 Oct 79	HS: 29-30 Apr 75
Ponchatoula (AO-148)	12 Jan 56	MUC: 8-22 Oct 68
30 Sep 66-26 Mar 73 (54 tours)	5 Sep 80	

Constructed in the mid-1950s, the six ships of the *Neosho* class were the first oilers built after World War II, and the first expressly designed as naval oilers rather than conversions of civilian tanker designs. At over 650 feet in length, with a capacity of 180,000 barrels of fuel, the *Neosho*s were larger than any previous USN oilers (including the jumboized *Cimarron*s). Fast, and able to deliver much product to fleet units before returning to port to replenish their liquid cargo, the three ships that served in Vietnam were in great demand.[9]

Photo 19-8

USS *Ponchatoula* (AO-148) under way off Oahu, Hawaii, 26 March 1970.
Naval History and Heritage Command photograph #NH 98832

ROUTINE AND UNCOMMON OILER OPERATIONS

Every ship in the Navy, except those nuclear-powered, require large quantities of diesel fuel, and those propelled by split atoms need diesel and jet fuel for the boats and aircraft they carry on board. Thus, oilers were critical to every type of operation, from naval gunfire support off the coast, to amphibious operations along the coast, to the recovery of Apollo astronauts in the mid-Pacific. Servicing customer ships at sea sometime involved battling typhoons while en route to rendezvous points, and occasional encounters with Soviet intelligence-gathering ships once engaged in refueling.

This chapter closes with overviews of the routine operations of *Ponchatoula* (AO-148) during her 1971-1972 deployment; the support provided by *Chipola* (AO-63) and *Hassayampa* (AO-145) during recovery of manned space capsules in the Pacific; and the 1969 deployment of *Ashtabula* (AO-51), her first after being jumboized. During her time, *Ashtabula* overcame much adversity while steadfastly carrying out her mission. She earned a Meritorious Unit Commendation in the process—one of twenty-five such awarded to AOs in Vietnam.

PONCHATOULA'S 1971-1972 DEPLOYMENT

On 22 July 1971, *Ponchatoula* stood out of Pearl Harbor on yet another deployment to Vietnam. She had begun her first line swing in the war zone on 30 September 1966; her last would end on 26 March 1973. On this deployment, she completed nine of her eventual fifty-four total swings. Following the last of the nine, and a 10-day period at Subic for provisioning and maintenance, she began the return transit to Pearl Harbor, arriving at her homeport on 2 February 1972.[10]

As evidenced by her record of operations, Subic was *Ponchatoula*'s home away from home while deployed; where she replenished fuel and stores, and her crew rested up between swings.

USS *Ponchatoula* 1971-1972 Deployment Schedule

Subic Bay (50 days)	Market Time and/or Yankee Station (100 days)	Non-Subic Liberty (17 days)
	7-18 Aug: Market Time	
19-22 Aug: Subic		
	23-25 Aug: Yankee Station	
26 Aug-6 Sep: Subic		
		7-13 Sep: Hong Kong
	14-28 Sep: Yankee Station	
29 Sep-4 Oct: Subic		
	5-26 Oct: Evading Typhoon HESTER and Yankee Station	
		27-29 Oct: Kaohsiung
30 Oct-5 Nov: Subic		
	5-13 Nov: Yankee Station	
14-19 Nov: Subic		
	20 Nov-7 Dec: Market Time and Yankee Station	
8 Dec: Subic		
	9-14 Dec: Yankee Station	
14 Dec: Subic		
	15-22 Dec: Yankee Station	
		23-29 Dec: Hong Kong
30 Dec-1 Jan: Subic		
	2-9 Jan: Yankee Station	
10-19 Jan: Subic[11]		

Photo 19-9

USS *Ponchatoula*, and other Service Force ships, alternated swings on the line with downtime at Naval Station, Subic Bay, for replenishment, maintenance, and crew rest, before their next period of high tempo operations off Vietnam. USS *Blakely* (DE-1072) Western Pacific 1973 cruise book (top photo); USS *Ponchatoula* (AO-148) Western Pacific 1971-1972 cruise book (other photos)

SUPPORT FOR MANNED SPACECRAFT RECOVERY

Photo 19-10

Piece of "philatelic mail" marked aboard USS *Chipola* with a specially designed cachet and stamped 27 December 1968, as a souvenir of the Apollo 8 manned flight.

Three fleet oilers were a part of Task Force 130 (Manned Spacecraft Recovery Force, Pacific) during the Apollo 8 and Apollo 11 missions, previously described in Chapter 11. Pearl Harbor was the headquarters of the Pacific Fleet and a hub for the recovery of Apollo space capsules that "splashed down" in the Pacific. The ships of Task Force 130 under Rear Adm. Fred E. Bakutis, USN, sailed from Pearl Harbor to meet the returning capsules, and take their crews back to Honolulu.[11]

Apollo 8 (21-27 December 1968)
- *Chipola* (AO-63)
- *Chuckawan* (AO-100)

Apollo 11 (16-24 July 1969)
- *Hassayampa* (AO-145)

Bakutis was both commander of the Hawaiian Sea Frontier, and commandant of the Fourteenth Naval District, with additional duty as commander Fleet Air, Hawaii. In his role as commander, TF 130, Bakutis was responsible for planning, implementing and coordinating all aspects of recovery operations in support of Manned Spacecraft programs in the Pacific Command Area.[12]

Astronauts returning from space were quarantined immediately after splashdown in a Mobile Quarantine Facility (MQF) aboard the principal recovery ship as it sailed back to Pearl Harbor. There, the MQF with the crew inside was taken off the ship, transported by land to Hickam Air Force Base, and placed aboard a C-141 aircraft for flight back to the Apollo Mission Control Center in Houston, Texas.[13]

For their support of the Apollo 8 space capsule recovery on 16 and 21 December 1968, respectively, *Chipola* and *Chuckawan* each received a Meritorious Unit Commendation. *Hassayampa* received a MUC for her participation with the Apollo 11 recovery force from 20-22 July 1969.

ASHTABULA'S FIRST JUMBOIZED DEPLOYMENT

> *While the "new"* ASHTABULA *celebrated her first birthday by holding swim call off the China mainland, we recalled her history during the last quarter century which saw this ship actually at sea more than 17 of her 25 years and we couldn't help think that for such an "old" ship she did pretty well. Sometimes we felt like we wanted to scuttle the ship when it seemed everything was breaking down (always in the middle of the night), but in the final analysis it is always the people that make up a ship's personality and it's the crew that dictates the final performance.*
>
> —USS *Ashtabula* (AO-51) Western Pacific 1969 cruise book

In the summer of 1969, while *Hassayampa* (back at Pearl Harbor from deployment to Vietnam, where she'd completed her last swing on 17 April) was participating in the Apollo space capsule recovery, the jumboized fleet oiler *Ashtabula* was battling Typhoon TESS. During her transit on 10 July, returning to Subic Bay after refueling ships involved in MARKET TIME operations off the Vietnam coast, she ran into TESS. *Ashtabula* attempted to evade the storm by turning south, but the center turned into her path, and the oiler found herself facing 25-foot swells and 125 mph winds. Fortunately, damage to the ship was minimal, despite coming within five miles of the eye, and enduring the worse part of the typhoon for four hours.[14]

While USS *Ponchatoula*'s deployment to Vietnam in 1971-1972 was typical of oiler duty, *Ashtabula*'s in 1969 included many challenges rolled into a single cruise. Photographs of her near the eye of a typhoon; an encounter with a Soviet intelligence-gathering ship; and a portrait of Barbara Eden, addressed by her to the crew of the *Ashtabula* follow.

Photo 19-11

Fleet oiler USS *Ashtabula* encountering Typhoon TESS on 10 July 1969, during transit to Subic Bay, following a period on the line refueling ships off Vietnam. USS *Ashtabula* (AO-51) Western Pacific 1969 cruise book

Not all vessels approaching *Ashtabula* were welcome. Soviet AGIs (intelligence-gathering ships) often endeavored to harass Navy ships replenishing at sea, and gather intelligence by photographing them and recording their electronic emissions. As detailed in Chapter 14, fleet tugs or salvage ships not engaged in other activities were often tasked with shouldering away such intruders. In the top photo, below, the fleet tug USS *Cree* (ATF-84) is keeping a trawler at bay in the Tonkin Gulf. The carrier USS *Oriskany* (CVA-34) is in the foreground, on *Ashtabula*'s port side. In the bottom photo, the tug continues these duties as the fleet oiler refuels a destroyer along her starboard side.

Photo 19-12

Fleet tug USS *Cree* (ATF-84) prevents a Soviet intelligence-gathering ship from getting any closer to a group of Navy ships replenishing at sea.
USS *Ashtabula* (AO-51) Western Pacific 1969 cruise book

SUPPORT FOR THE TROOPS BY ENTERTAINERS

Several American actresses, some associated with Bob Hope USO tours, others acting independently, visited the war zone during the Vietnam War to entertain the troops. Others helped to keep morale up (like "pinups" in previous wars) by making available portraits of themselves upon request. Barbara Eden—an American film, stage, and television actress, and singer, best known for her starring role in the American TV sitcom, "I Dream of Jeannie," which aired from 1965-1970—did so for the crew of USS *Astabula*. (Viewing the comely actress in a swim suit, probably helped make the long days at sea a little easier, for at least some crewmembers.) Two decades later, she joined Bob Hope's 1987 USO Christmas Show entertaining U.S. Navy personnel helping protect oil tankers and other ships in the Persian Gulf from constant, unprovoked attacks during the Iran-Iraq War.

Photo 19-13

Barbara Eden in the 1960s, and with Bob Hope in 1987, while entertaining sailors and Marines aboard USS *Okinawa* (LPH-3) at anchor off Bahrain Bell in the Persian Gulf. (The author was at the performance that night.)
Photo on the left is from USS *Ashtabula* (AO-51) Western Pacific 1969 cruise book; the one on the right was taken by PH2(SW) Jeff Elliot

20

Fast Combat Support Ships and Replenishment Oilers

To meet her destiny, to the Tonkin she came
And with sweat and guts she brought her fame
To CAMDEN sailors there was no day or night
Just "Man both sides," for the carriers [are] in sight

Though our guns never fire a lethal round
And we've never been deafened by battle's sound
And we've never tracked MIGs over NVN skies
We've fought the war with the rest of the guys

—Two stanzas of the poem "The Legend called CAMDEN" from USS *Camden* (AOE-2) Western Pacific 1968-1969 cruise book

Photo 20-1

The attack carrier *Hancock* (CVA-19) (at right) and the guided missile destroyer *Robison* (DDG-12) (at left) refuel from the fast combat support ship *Sacramento* (AOE-1) (center) during operations in the Western Pacific, 1 October 1965. National Archives photograph #USN 1109945

The Navy's two AOEs (fast combat support ships) and five AORs (replenishment oilers) that served in Vietnam were all new ships. The first of them, *Sacramento*, was commissioned in 1964, and the last, *Wabash*, in 1971. Summary information about the AOEs and AORs, including their span of service in Vietnam, from the start of their first tour to the end of their last tour, and unit awards they received, follow.

Fast Combat Support Ships (AOE)
793 feet; 49,956 tons; 26 kts; 22 officers, 530 enlisted

Replenishment Oilers (AOR)
659 feet; 39,790 tons; 20 kts; 34 officers, 463 enlisted

Ship Name, Ship Class, Span of Vietnam Service	Comm/ Decom	Unit Commendations
Camden (AOE-2) *Sacramento*-class 2 Jul 68-25 Aug 72 (28 tours)	1 Apr 67 29 Sep 05	MUC: 26 Jun 68-22 Feb 69 MUC: 6 Sep 70-12 Mar 71 NUC: 14 Feb-7 Sep 72
Sacramento (AOE-1) *Sacramento*-class 6 Nov 65-13 Mar 73 (46 tours)	14 Mar 64 1 Oct 04	MUC: 15 Mar-18 Aug 71
Kansas City (AOR-3) *Wichita*-class 12 Jun 71-26 Nov 72 (13 tours)	6 Jun 70 7 Oct 94	MUC: 25 Apr-12 Dec 72
Milwaukee (AOR-2) *Wichita*-class 12 Nov 72-20 Feb 73 (5 tours)	1 Nov 69 27 Jan 94	
Savannah (AOR-4) *Wichita*-class 27 May 72-1 Nov 72 (6 tours)	5 Dec 70 28 Jul 95	MUC: 15 May-15 Nov 72
Wabash (AOR-5) *Wichita*-class 9 Dec 72-28 Mar 73 (5 tours)	20 Nov 71 30 Sep 94	
Wichita (AOR-1) *Wichita*-class 21 Jul 70-18 Feb 73 (23 tours)	7 Jun 69 12 Mar 93	MUC: 4 Jul 70-19 Jan 71 MUC: 29 Jul 72-5 Mar 73

SACRAMENTO'S LENGTHY WAR SERVICE

Sacramento, commissioned on 14 March 1964 (the first of the Navy's four *Sacramento*-class fast combat support ships), was propelled by two of the four steam turbines manufactured for the battleship *Kentucky* (BB-66), which was never completed. Her cargo capacity was 175,000 barrels or nearly eight million gallons of Navy Standard fuel oil; 1,600 tons of ammunition, including missiles; and 250 tons of dry stores and refrigerated provisions. Aviation facilities include a loading area aft and hangar space for three UH-46 helicopters.[1]

Sacramento departed the Bremerton/Seattle area on 14 October 1965, bound for the Western Pacific and duty in Vietnam. Arriving at Subic Bay on the 28th, she joined Task Force 73, the Seventh Fleet replenishment and mobile logistics support force. She then proceeded to Yankee Station off Vietnam's coast to take up replenishment duties. A typical cycle consisted of two and one-half weeks on Yankee Station, a quick run back to Subic Bay for five or six days of in-port loading, then a return to station. *Sacramento*'s nine-month deployment ended with her arrival at Seattle on 17 July 1966.[2]

Photo 20-2

POL (Petroleum, oil and lubricants) Pier at Subic, where *Sacramento* loaded same. USS *Sacramento* (AOE-1) Western Pacific 1966-1967 cruise book

Sacramento arrived back in the South China Sea in early December 1967. Customers of the Tonkin Gulf "seagoing supermarket" were many, and diverse—including carriers, amphibs, cruisers and destroyers, salvage ships, and at least one ocean going minesweeper. The sweep had to make nearly flank speed to maintain station in heavy seas alongside the much faster AOE. (Minesweeper engines are designed to produce torque necessary to tow heavy sweep gear, at the expense of higher speed desired in other types of ships.)[3]

Photo 20-3

Top: A carrier approaches *Sacramento*'s port side to take up her refueling station.
Bottom: An ocean minesweeper receives fuel from the *Sacramento* in heavy seas.
USS *Sacramento* (AOE-1) Western Pacific 1966-1967 cruise book

The work of UnRep (underway replenishment) alongside and VertRep (vertical replenishment) via helicopter was seemingly never ending, and took place at any time, day or night. Of the 583 ships *Sacramento* serviced during her seven-month deployment, 370 were unreps and 213 were vertreps. The embarked helicopter detachment logged 559 flight hours during ship-to-ship transfers.

Photo 20-4

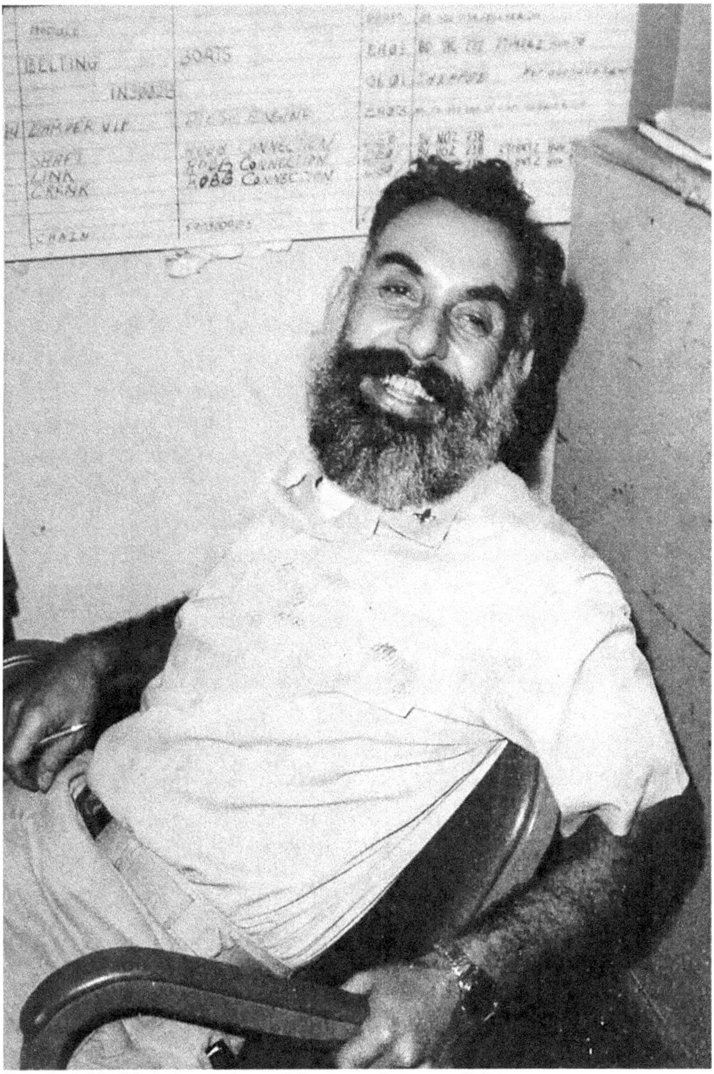

Bearded, khaki-clad member of *Sacramento*'s cargo division.
USS *Sacramento* (AOE-1) Western Pacific 1970 cruise book

This pattern of service continued throughout the war; *Sacramento* departing America's west coast on deployment, returning home after several months' service off Vietnam and, following maintenance and training, deploying once again. *Sacramento*'s next Western Pacific cruise began on 6 January 1968. After replenishing the 351st ship of that deployment on 15 June, she set a course for the West Coast. From San Francisco, the stalwart AOE proceeded north to Bremerton, Washington, for her first regular yard overhaul since commissioning.[4]

On 11 February 1969, *Sacramento* again steamed west, arrived at Subic Bay on the 28th, and five days later headed for Yankee Station. She completed 471 replenishments over the course of this deployment, and returned home at the end of September.[5]

Sacramento departed on her fifth Western Pacific deployment for duty in Vietnam on 23 February 1970, after loading ammunition at Bangor and fuel at Manchester—located on the Kitsap Peninsula on the shores of Puget Sound, across the Sound from Seattle. She arrived in Subic Bay on 15 March and six days later, headed for Yankee Station. It wasn't unusual during this cruise for eight to ten ships to come alongside in a day for fuel, ammo, stores, and mail. After arriving in San Diego on 18 September, at deployment's end, *Sacramento* sailed up the West Coast to Puget Sound, Washington, for ammunition offload, a stand-down period (crew leave, generous liberty, and reduced shipboard work), and a restricted availability.[6]

Two more deployments to Vietnam awaited *Sacramento*. The AOE sailed west from Bangor, Washington, on 1 March 1971 and entered Subic Bay on the 20th. She stood out again four days later to take up her first "line swing" of this tour. *Sacramento* scuttled back and forth between Subic Bay and Yankee Station until 5 August, when she set a course for Sasebo, Japan, on the first leg of her voyage back to the United States.[7]

She sailed from Bangor on 11 August 1972, en route to the war zone for the final time. *Sacramento* replenished units of the Seventh Fleet for the next six and one-half months, finally departing Subic Bay on 23 March 1973. After visiting Yokosuka, she began the Pacific crossing, arriving at Bremerton on 16 April.[8]

SECOND AOE BEGINS DEPLOYMENTS TO VIETNAM

THE CAMDEN WAY. We're Number Two—We Tryer Harder!

—From USS *Camden* (AOE-2) Western Pacific 1970-71 cruise book

USS *Camden* (AOE-2), sister ship to the *Sacramento*, was commissioned on 1 April 1967. She too was propelled by battleship turbines, having the other two originally intended for the *Kentucky* (BB-66). After it was decided to scrap the uncompleted battleship, the Navy divided the propulsion system between *Sacramento* and *Camden*.[9]

Constructed at New York Shipbuilding Co., Camden, New Jersey, AOE-2 joined the Pacific Fleet in September 1967, which required passage through the Panama Canal. She made her maiden overseas deployment from May 1968 through March 1969, and was awarded a Meritorious Unit Commendation for her efforts.[10]

Photo 20-5

In the Gulf of Tonkin, 8 February 1969, the destroyer USS *Carpenter* (DD-825) pulls alongside the *Camden* to take on supplies by highline and helicopter. At the same time, the carrier USS *Hancock* (CVA-19) receives fuel from *Camden*.
National Archives photograph #USN 1137949

Camden left Long Beach, California, on 21 August 1969 on her second deployment and returned to homeport on 26 March 1970, having serviced 356 Seventh Fleet ships while in the Pacific. For her third deployment covering the period 26 August 1970 to April 1971, *Camden* was awarded a second Meritorious Unit Commendation. She departed Long Beach on 4 February 1972, beginning her fourth and final Vietnam deployment. She arrived home on 15 September, and received a Navy Unit Commendation for that cruise—making her the most decorated of the AOEs/AORs that served in Vietnam.[11]

AORS DEPLOY TO VIETNAM IN 1970, 71, 72, AND 73

USS *Wichita* (AOR-1)—the first ship of a class of seven replenishment oilers—was commissioned on 7 June 1969. The *Wichita*-class was based on the 796-foot *Sacramento*, the first ship in a class of AOEs specifically designed to provide logistical support for an attack carrier battle group. The original concept was that AORs would serve the same function for CVS carrier groups—centered around smaller aircraft carriers, whose primary mission was anti-submarine warfare—as did the larger, faster AOEs for the attack carrier (CVA) groups. *Wichita* was two-thirds the size of *Sacramento*, slower, and carried less cargo. On the plus side, she was less expensive to both procure and operate, as a result of her economical design, lower construction cost, and smaller crew.[12]

Wichita stretched 658 feet in length, spanned 96 feet in breadth, and drew 33 feet of water. Her profile featured a raked bow, battleship stern, and two islands, including a forward conning station and an aft deckhouse. She displaced 13,662 tons, and her propulsion plant, generating 32,000 shaft horsepower could propel the AOR at 20 plus knots. *Wichita* and the other replenishment oilers could carry 160,000 barrels of fuel, 600 tons of munitions, 200 tons of dry stores and 100 tons of refrigerated stores.[13]

Photo 20-6

USS *Wichita* in heavy seas, location and date unknown.
USS *Wichita* (AOR-1) Western Pacific 1970-1971 cruise book

Five AORs served in Vietnam between 21 July 1970 (the beginning of *Wichita*'s first tour on the line) and 28 March 1973 (the end of *Wabash*'s final tour). *Wichita* did twenty-three swings on the line between 21 July 1970 and 18 February 1973. The other replenishment oilers—*Kansas City*, *Milwaukee*, *Savannah*, and *Wabash*—collectively completed another twenty-nine swings (tours in the combat zone).

The following summary information related to *Wichita*'s 1970-1971 deployment, evidences the vast amount of work carried out by her "Wichita linemen." This crew moniker was likely adopted from the title of a song, "Wichita Lineman," recorded by American country music artist Glen Campbell in 1968. His recording reached number three on the U.S. pop chart, and topped the country music chart for two weeks.

USS *Wichita* (AOR-1) 1970-1971 Deployment

Line Swing	Duration	#Ships Served	Fuel Gallons	Ammo Tons	Provisions Tons
1	21 days	48	7,211,400	2,812.53	177.7
2	33 days	42	4,846,800	1,510.33	48.8
3	23 days	48	8,862,000	1,565.78	169.6
4	47 days	124	16,888,200	2,672.97	361.8
5	9 days	13	2,490,600	585.67	25.2
6	9 days	45	8,988,000	3,177.51	28.3
7	6 days	16	3,019,800	629.04	16.0
Total	148 days	328 [sic]	52,306,800	13,246.83 [sic]	827.4[14]

Note: A total of 336 ships came alongside *Wichita* and she transferred a total of 12,953.83 tons of ammunition.

On her first deployment from Long Beach to the Western Pacific, Wichita joined the Seventh Fleet on 4 July 1970, calling at Subic Bay before proceeding off the coast of Vietnam. In between multiple tours on the line, she visited Hong Kong twice, and enjoyed liberty in the Philippines, while loading out for her next "swing." (This term referred to a period of duty, commencing upon leaving Subic for Vietnam, and ending back in Subic to load for another tour in the combat zone.) The replenishment oiler arrived home in Long Beach on 2 February 1971.[15]

Wichita deployed from Long Beach on 7 August 1971, and arrived at Subic Bay on 24 August. She began provisioning the battle group operating off Vietnam at month's end. Her routine was broken toward the end of October. First, by a visit to Sattahip, Thailand, and then assignment on 10 December, to join a task force heading toward the Indian Ocean, to observe the Indo-Pakistani War and react if necessary. *Wichita* arrived back off Vietnam in early January 1972, and resumed provisioning the battle group. She continued supporting the fleet until her departure in February, bound for Long Beach. Following arrival there on 31 March, she made her way up the coast to San Francisco, for overhaul at Hunters Point Naval Shipyard.[16]

Wichita arrived once again at Subic Bay on 4 August 1972, at the beginning of her third deployment. After stocking her holds, she took up replenishment operations. Multiple trips back and forth between the battle group and Subic Bay followed. Ship's force enjoyed liberty at Hong Kong and Sattahip during the cruise. *Wichita* returned to Long Beach on 16 March 1973, during the closing stages of American operations in the Vietnam War.[17]

NORTH VIETNAM'S EASTER OFFENSIVE

> *Every minute, hundreds of thousands of people die on this earth. The life or death of a hundred, a thousand, tens of thousands of human beings, even our compatriots, means little.*
>
> —Remark attributed to Vo Nguyen Giap after defeating the French at Dien Bien Phu in May 1954, forcing France from Indochina. The relentless and charismatic North Vietnamese general was from the early 1960s to the mid-1970s, perhaps second only to his mentor, Ho Chi Minh, as the face of a tenacious, implacable enemy.[18]

In early March 1972, the relatively calm state of the South Vietnamese countryside seemed to vindicate the Nixon administration's "Vietnamization" policy—withdrawal of U.S. ground combat forces from the country while also improving the capability of the armed forces of the Republic of Vietnam. Richard M. Nixon had succeeded Lyndon B. Johnson as president on 20 January 1969 amid much dissent and unrest throughout the nation, resulting from the long war that had taken many American lives. During his campaign, Nixon pledged to get the United States out of Vietnam and, after taking office, he directed additional actions to expand, equip, and train Vietnamese military forces, steadily reduce American troops, and give South Vietnam greater responsibility for fighting the war.[19]

This lull in warfare, however, would prove illusionary. In Hanoi, the senior North Vietnamese general, Vo Nguyen Giap, was planning a massive invasion designed to destroy the South Vietnamese armed forces and capture South Vietnam. Giap hoped for a conclusive outcome, or at least to seize enough territory for North Vietnam to improve its negotiating position in Paris. At noon on 30 March 1972, the North Vietnamese 308th Army Division plus two independent regiments struck the ARVN (Army of the Republic of Vietnam) fire support bases along the DMZ. Concurrently, the 304th Division rolled out of Laos striking past Khe Sanh toward Quang Tri City.[20]

The "Easter Offensive" mounted by the North Vietnamese Army, and led by Soviet-built tanks moving across the DMZ into South Vietnam's Quang Tri province, was the first major assault since the Tet Offensive in January 1968. The difference between the Tet '68 and Easter '72 offensives was that in 1968 the United States had been escalating its operations. In 1972, military action was phasing down and the North Vietnamese knew it. Most of the major U.S. ground forces had already left Vietnam, Nixon had just made his historic visit to Peking, and Secretary of State Henry Kissinger was in Paris negotiating with the North Vietnamese for an end to the Vietnam War.[21]

On 2 April, President Nixon ordered strikes against North Vietnam near the DMZ by air and sea. Task Force 77 increased in size to five carriers—*Constellation* (CVA-64), *Kitty Hawk* (CVA-63), *Hancock* (CVA-19), *Coral Sea* (CVA-43), and *Saratoga* (CVA-60)—the largest concentration of carriers in the Gulf of Tonkin during the war. The air squadrons hit key military and logistic facilities at Dong Hoi, Vinh, Thanh Hoa, Haiphong, and Hanoi. Aircraft also attacked enemy troop units, supply convoys, and headquarters in the areas around the DMZ. Also taking part were cruisers and destroyers, which ranged the southern North Vietnamese coastline, shelling enemy transportation routes, troop concentrations, shore defenses, and logistic installations.[22]

By the end of September 1972, the North Vietnamese diplomats in Paris were much more amenable to serious negotiation than they had been at the end of March. Believing that a negotiated settlement of the war was within reach in Paris, on 11 October the Nixon administration ordered U.S. Pacific forces to cease bombing in the vicinity of Hanoi. Twelve days later, Washington restricted allied strikes to targets below the 20th parallel. Nevertheless, negotiations again bogged down in Paris while the North Vietnamese strengthened the air defenses of the capital and Haiphong, restored rail lines to China, and stockpiled war reserves.[23]

In response to these actions, Nixon ordered a massive air assault by Air Force B-52 bombers, tactical aircraft, and the Navy's carrier aircraft against targets in Hanoi and Haiphong. The operational area was later broadened to include Thai Nguyen, Long Dun Kep, and Lang Dang as well. The American offensive had the desired effect. At year's end the

North Vietnamese resumed serious discussions in Paris, and both sides ceased combat operations in the north in mid-January 1973.[24]

RECOGNITION OF AOE AND AOR PERFORMANCE

Of the eight Unit Commendations awarded fast combat support ships and replenishment oilers during the war, four were for duty on the line during North Vietnam's Easter Offensive and/or resultant dramatically increased American offensive operations through the end of 1972.

Ship	Span of Service	Unit Commendations
Camden (AOE-2)	2 Jul 68-25 Aug 72	MUC: 26 Jun 68-22 Feb 69
		MUC: 6 Sep 70-12 Mar 71
		NUC: 14 Feb-7 Sep 72
Sacramento (AOE-1)	6 Nov 65-13 Mar 73	MUC: 15 Mar-18 Aug 71
Kansas City (AOR-3)	12 Jun 71-26 Nov 72	MUC: 25 Apr-12 Dec 72
Milwaukee (AOR-2)	12 Nov 72-20 Feb 72	
Savannah (AOR-4)	27 May 72-1 Nov 72	MUC: 15 May-15 Nov 72
Wabash (AOR-5)	9 Dec 72-28 Mar 73	
Wichita (AOR-1)	21 Jul 70-18 Feb 73	MUC: 4 Jul 70-19 Jan 71
		MUC: 29 Jul 72-5 Mar 73

21

War's End

Don't you people realize what's happening?... There is no longer a consensus of support for the war back in the United States. I have a letter in my pocket from the president [Lyndon B. Johnson] that tells me to turn the war over to the Vietnamese.... You tell me that we'll be all turned over by 1976. That's out of the question! The country will not sit still for that kind of commitment. The president wants to get the war turned over as soon as possible. We have to make that happen.

—Gen. Creighton W. Abrams Jr., commander, U.S. Military Assistance Command, Vietnam, addressing his chief advisors at a meeting on 2 November 1968.[1]

Photo 21-1

On 18 July 1973, the Seventh Fleet departed North Vietnamese territorial waters, following the clearance of carrier aircraft-laid mines in Haiphong Harbor (a provision for the return of American POWs), ending the U.S. Navy's involvement in Vietnam. USS *Maury* (AGS-16) and *Serrano* (AGS-24) Vietnam Survey 1965-1966 cruise book

On 18 July 1973, the Seventh Fleet departed North Vietnamese territorial waters, ending the U.S. Navy's long, arduous, and costly involvement in Vietnam. This effort was completed on 29 March 1973 with the departure of all remaining U.S. Navy and Marines forces, except embassy personnel, from Vietnam. The exodus of U.S. military forces from Vietnam fulfilled a promise made in 1968 by then presidential-candidate Richard Nixon. Nixon had succeeded President Lyndon Johnson on 20 January 1969, amid much dissent and unrest throughout the nation, resulting from a long war that had taken many American lives.[2]

During the latter part of 1968, dramatic changes occurred in the conduct of the war, as a result of the enemy's bloody country-wide Tet Offensive of February and March and the follow-up attacks during the spring. The Johnson administration, convinced that the allied military effort was faring badly, and in recognition of increasing domestic opposition to the American role in the war, ordered the gradual withdrawal of U.S. forces from Southeast Asia. At the same time, the first significant attempt at peace talks came in May 1968 with an informal meeting between U.S. and North Vietnamese envoys in Paris. Five months later, Lyndon Johnson agreed to suspend all bombing sorties over North Vietnam, paving the way for formal peace negotiations.[3]

Photo 21-2

Commencement of the Vietnam Peace talks on 25 January 1969. The U.S. delegation is in foreground with chief negotiator Henry Cabot Lodge, facing camera.
U.S. Information Agency photograph #69-526

On 25 January 1969, five days after the inauguration of Richard M. Nixon as the 37th President of the United States, negotiators from Washington flew to Paris for peace meetings with representatives of North and South Vietnam and the National Liberation Front. The Paris peace talks would drag on for more than four years, plagued by setbacks and breakdowns in negotiations. The North Vietnamese demanded the withdrawal of American troops, the dissolution of the South Vietnamese government, and a return to the principles of the Geneva Accords; while the United States insisted that Hanoi recognize the sovereignty of South Vietnam.[4]

During his campaign, Nixon had pledged to get the United States out of Vietnam. Once in office, he had directed additional actions to expand, equip, and train Vietnamese military forces, steadily reduce American fighting troops, and give South Vietnam greater responsibility for fighting the war. The American military termed its part of the administration's "Vietnamization" of the war, the Accelerated Turnover to the Vietnamese Program.[5]

As United States forces began to train and equip the South Vietnamese military to assume complete responsibility for the war, they also worked to keep pressure on the enemy. From 1968 to 1971, the allies exploited the Communists' staggering losses during the Tet attacks by pushing the enemy's large main force units out to the border areas, thereby extending Allied presence into Viet Cong strongholds, and consolidating control over population centers.[6]

As discussed in the preceding chapter, a lull in warfare, which seemed to vindicate the Nixon administration's withdrawal of U.S. ground combat forces from the country while also improving the capability of the armed forces of the Republic of Vietnam, would prove illusionary. In 1972, military action was phasing down and the North Vietnamese knew it. Most of the major U.S. ground forces had already left Vietnam, Nixon had just made his historic visit to Peking, and Secretary of State Henry Kissinger was in Paris negotiating with the North Vietnamese for an end to the Vietnam War.[7]

NIXON ORDERS AIR AND MINING CAMPAIGN

> *I have determined that we should go for broke.... We must punish the enemy.*
>
> —President Richard M. Nixon expressing in a memorandum to Secretary of State Henry Kissinger, his desire to resume a broad air interdiction campaign against North Vietnam, and a naval blockade of the entire North Vietnamese coast, including mining actions against Haiphong and other major harbors.[8]

North Vietnam's "Easter Offensive" (invasion of South Vietnam on 30 March 1972) violated the Geneva agreements and justified strong retaliatory action by the United States. President Nixon favored two responses: the resumption of a broad air interdiction campaign against North Vietnam, and a naval blockade of the entire North Vietnamese coast, including the mining of Haiphong and other major harbors. The latter action was intended to starve North Vietnam of war materiel and supplies which had been arriving regularly at its ports up to this point in the protracted war, aboard Chinese and Soviet merchant shipping.[9]

Photo 21-3

Jacket patches commemorating the actions of minesweepers, minesweeping helicopters, and support ships of Task Force 78 in the clearance of mines from Haiphong Harbor—a condition for the release of American prisoners of war by North Vietnam.

Extensive bombing and shore bombardment of a broad array of enemy targets begun on 9 May 1972, in concert with the mining of Haiphong Harbor, was instrumental in bringing the North Vietnamese back to the negotiating table in the Paris cease-fire talks that August. On 27 January 1973, U.S., South Vietnamese, North Vietnamese, and Viet Cong representatives finally signed a cease-fire agreement at Paris. Under its provisions, the Communists agreed to release all American prisoners of war within a space of two months in exchange for U.S. military withdrawal from South Vietnam and the U.S. Navy's clearance of mines from North Vietnamese waters.[10]

SERVICE FORCE SUPPORTS NAVAL OPERATIONS IN NORTH VIETNAMESE WATERS INTO MARCH 1973

Twenty-eight Service Forces ships—10 AOs, 1 AOR, 9 AEs, 1 AOE, 3 AFS, 3 ATF, and 1 ARS—continued to serve "on the line" into the early months of 1973. On 28 March, the final five ships left the war zone off Vietnam for the final time. Only the period of the ships' last tour (swing) is provided in the table. Many of them had preceding swings earlier in 1973 and, in some cases, in late 1972.

Some of the ships operated near the North Vietnamese coast in direct support of Rear Adm. Brian McCauley's Task Force 78, engaged in mine clearance operations. Others were in the Tonkin Gulf, caring for the aircraft carriers and surface combatants standing by, should any difficulties arise during the closing stages of the war.

January 1973
(5 ammunition ships, 4 fleet oilers)

Ship	Last "Line Swing"
Waccamaw (AO-109)	28 Dec 72-5 Jan 73
Santa Barbara (AE-28)	6-18 Jan 73
Mauna Kea (AE-22)	1-25 Jan 73
Hassayampa (AO-145)	10-25 Jan 73
Mount Katmai (AE-16)	13-27 Jan 73
Vesuvius (AE-15)	17-29 Jan 73
Nitro (AE-23)	26-29 Jan 73

February 1973
(1 ammunition ship, 2 fleet oilers, 1 replenishment oiler)

Ship	Last "Line Swing"
Suribachi (AE-21)	26 Jan-8 Feb 73
Passumpsic (AO-107)	6-12 Feb 73
Wichita (AOR-1)	13-18 Feb 73
Manatee (AO-58)	14-18 Feb 73

March 1973
(3 ammunition ships, 5 fleet oilers, 1 replenishment oiler, 1 fast combat support ship, 3 combat stores ships, 3 fleet tugs, 1 salvage ship)

Butte (AE-27)	22 Feb-3 Mar 73
Caliente (AO-53)	2-10 Mar 73
Sacramento (AOE-1)	2-13 Mar 73
Mount Hood (AE 29)	5-13 Mar 73
Tolovana (AO-64)	10-16 Mar 73
Niagara Falls (AFS-3)	11-19 Mar 73
Marias (AO-57)	12-20 Mar 73
Chowanoc (ATF-100)	21 Feb-21 Mar 73
Flint (AE-32)	28 Feb-22 Mar 73
Cacapon (AO-52)	7-22 Mar 73
San Jose (AFS-7)	21-23 Mar 73
Ponchatoula (AO-148)	16-26 Mar 73
Tawasa (ATF-92)	21 Feb-28 Mar 73
Safeguard (ARS-25)	3-28 Mar 73
Moctobi (ATF-105)	21-28 Mar 73
White Plains (AFS-4)	24-28 Mar 73
Wabash (AOR-5)	25-28 Mar 73

AE:	Ammunition ship	AF:	Stores ship
AO:	Fleet oiler	AFS:	Combat stores ship
AOE:	Fast combat support ship	ARS:	Salvage ship
AOR:	Replenishment oiler	ATF:	Fleet tug

SERVICE FORCE PERSONNEL CASUALTIES

Although not addressed in this book, Service Force ships suffered some personnel losses during the Vietnam War, owing to combat action, accidents, and perhaps natural causes. Appendix I identies seventeen officers and men who did not return home.

Postscript

Support for the Fleet closes with a pictorial tour of common shipboard events and evolutions, followed by leisure activities sailors commonly enjoy aboard ship and ashore. A few of the photographs depict sailors desperately trying to get some needed rest or, having done so, engaged in athletic activities aboard ship. Others provide insight into an activity involving fun only for some. The remaining participants most likely wanted their involvement in a longstanding tradition among mariners—a Line Crossing Ceremony for those who had not previously sailed across the Equator—to end as soon as possible.

The bulk of the photographs are intended to help former sailors to reminisce about runs ashore with shipmates, and provide opportunity for those who were unable to visit one or more of the liberty ports depicted, to do so vicariously.

SHIPBOARD EVENTS AND ACTIVITIES

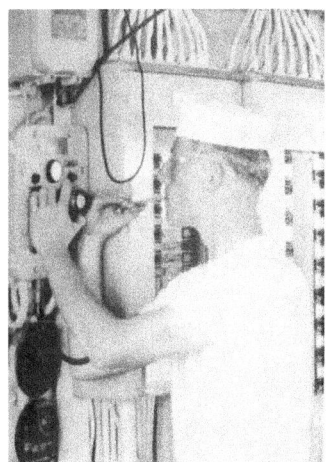 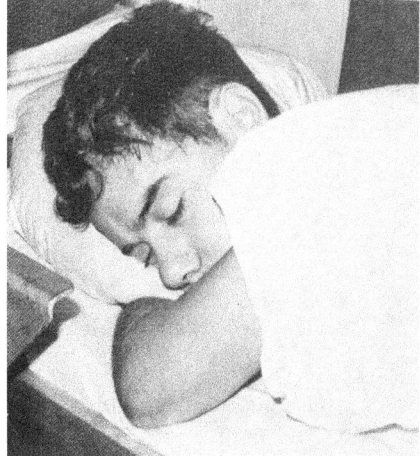

Aboard the fleet oiler *Ashtabula*, the boatswain's mate of the watch pipes reveille, while a crewmember tries to ignore the inevitable, and catch a few extra winks. USS *Ashtabula* (AO-51) Western Pacific 1969 cruise book

Small arms practice off the fantail of the repair ship USS *Markab* while under way.
USS *Markab* (AR-23) Western Pacific 1967 cruise book

S-1 Division stands patiently at parade rest awaiting Captain's Inspection.
USS *Castor* (AKS-1) 1967 cruise book

CROSSING THE LINE ("SHELLBACK INITIATION")

Twice in the cruise the good ship MATTAPONI *crossed into the domain of Neptunus Rex, and twice he consented to visit us with his Royal Court, first between Singapore and Kaohsiung, and again between Sasebo and Brisbane. He inspected the ship and found it seaworthy save the hordes of Pollywog Scum found slithering upon its decks. But by the time he left, the ship was once again pure and he could with good conscience let it proceed. But once the crew had all been initiated into the solemn mysteries of the deep, the only mystery remaining was how to clean all that garbage off the tank deck.*

—USS *Mattaponi* Western Pacific 1968-1969 cruise book, describing two interludes, not fun for many, from duty on the line.

A Slimy Pollywog working hard to become a Trusty Shellback.
USS *Mattaponi* (AO-41) Western Pacific 1968-1969 cruise book

A pollywog on another ship drags his slimy body through the Royal Tunnel of Love.
USS *Tappahannock* (AO-43) Western Pacific 1966-1967 cruise book

USS *Arlington*'s Royal Court prepares to oversee a Crossing the Line Ceremony.
USS *Arlington* (AGMR-2) Middle Pacific 1969 cruise book

Postscript 229

"Shellback" Initiation aboard *Annapolis*.
USS *Annapolis* (AGMR-1) Western Pacific 1967 cruise book

Royal Medicine.
USS *Arlington* (AGMR-2) Middle Pacific 1969 cruise book

Dunking tank and emergence of new Shellbacks.
USS *Arlington* (AGMR-2) Middle Pacific 1969 cruise book

Coffin for special cases.
USS *Arlington* (AGMR-2) 1968 cruise book

PRANK ON NEW, UNWARY SAILORS

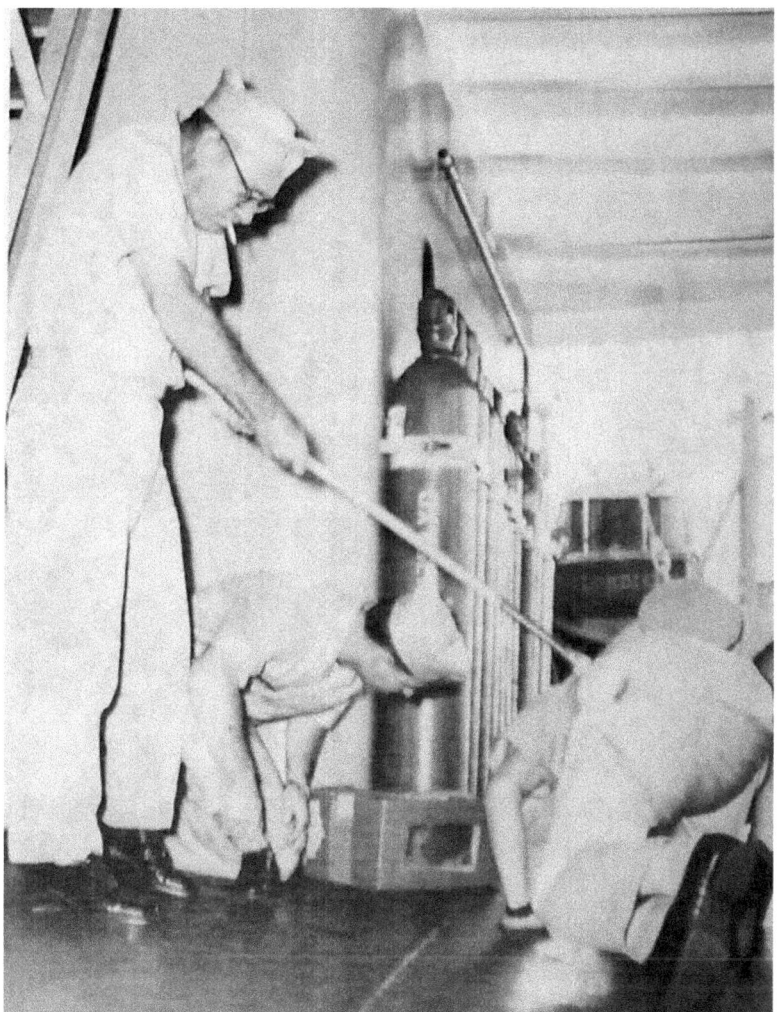

Newly reported crewmembers bending over to look at a captured "seabat" in the box. In this prank (a favorite in the fleet for decades), more than one victim, overcome by eagerness, has exclaimed, "Stop swatting me, I'm trying to see the seabat."
USS *Klondike* (AR-22) Western Pacific 1969 cruise book

ATHLETICS ABOARD SHIP

"Smokers" (boxing match) aboard USS *Arlington*.
USS *Arlington* (AGMR-2) 1968 cruise book

Postscript

Reaching the water when swim call was announced aboard the repair ship USS *Delta* involved a high dive off her deck, or slower transit via Jacobs ladder or boarding net. USS *Delta* (AR-9) Western Pacific 1969 cruise book

LEISURETIME ACTIVITIES

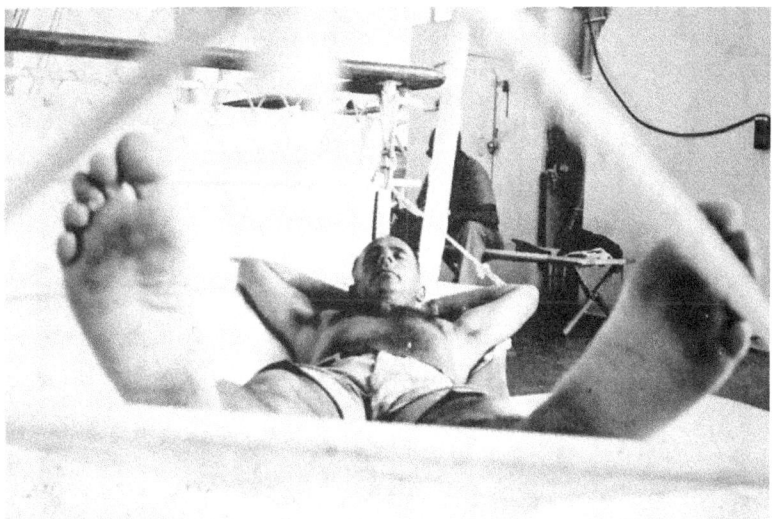

Crewmember getting some rest aboard the hospital ship *Sanctuary*. USS *Sanctuary* (AH-17) 1967 cruise book

Crewmembers enjoying time off work and watches aboard USS *Arlington*. USS *Arlington* (AGMR-2) 1968 cruise book

HONG KONG, BRITISH CROWN COLONY

A Hong Kong junk.
USS *Mars* (AFS-1) 1966 cruise book

Views of Hong Kong and its harbor by day and night.
USS *Arlington* (AGMR-2) 1968 cruise book

Postscript 237

Sailors could ride the ferry from anchorage for two cents and, once ashore, the China Fleet Club offered bargains galore for sailors with limited funds. USS *Castor* (AKS-1) 1967 cruise book

Rickshaw drivers taking a breather between paid customers, and sailors on liberty enjoying this form of transportation, after employing such services.
USS *Arlington* (AGMR-2) 1968 cruise book

Postscript 239

Other views of Hong Kong.
USS *Arlington* (AGMR-2) 1968 cruise book

People of Hong Kong.
USS *Arlington* (AGMR-2) 1968 cruise book

Postscript 241

JAPAN

Japanese woman in traditional dress.
USS *Arlington* (AGMR-2) 1968 cruise book

Beautiful Japanese architecture.
USS *Arlington* (AGMR-2) 1968 cruise book

Japanese artist at work.
USS *Arlington* (AGMR-2) 1968 cruise book

Postscript 243

Carved figure adorns a structure corner just below the roofline.
USS *Arlington* (AGMR-2) 1968 cruise book

Picturesque view of famed Mt. Fuji.
USS *Arlington* (AGMR-2) 1968 cruise book

Nagasaki Peace Statue, Nagasaki, Japan
USS *Ashtabula* (AO-51) Western Pacific 1969 cruise book

Tokyo's Ginza offers a shopping mecca by day, and a brilliantly lighted entertainment center by night.
USS *Arlington* (AGMR-2) 1968 cruise book

Daytime view of other nightspots frequented by sailors.
USS *Arlington* (AGMR-2) 1968 cruise book

"Thieves Alley," Yokosuka
USS *Niagara Falls* (AFS-3) Western Pacific 1968 cruise book

Flagship at Yokosuka used in the Russo-Japanese War.
USS *Niagara Falls* (AFS-3) Western Pacific 1968 cruise book

Sasebo, Japan
USS *Delta* (AR-9) Western Pacific 1969 cruise book

TAIWAN

A pretty girl.
USS *Arlington* (AGMR-2) 1968 cruise book

Street scene.
USS *Arlington* (AGMR-2) 1968 cruise book

250   Postscript

View overlooking the city and port of Kaohsiung, Taiwan.
USS *Castor* (AKS-1) 1967 cruise book

A traditional gate.
USS *Arlington* (AGMR-2) 1968 cruise book

THAILAND

Bangkok, Thailand.
USS *Castor* (AKS-1) 1967 cruise book

PHILIPPINE ISLANDS

National beer of the Philippines.
USS *Arlington* (AGMR-2) 1968 cruise book

Jeepneys providing public transportation crowd the streets.
USS *Arlington* (AGMR-2) 1968 cruise book

Postscript 253

Opportunity to convert dollars to pesos before heading out on liberty in Olongapo.
USS *Markab* (AR-23) Western Pacific 1967 cruise book

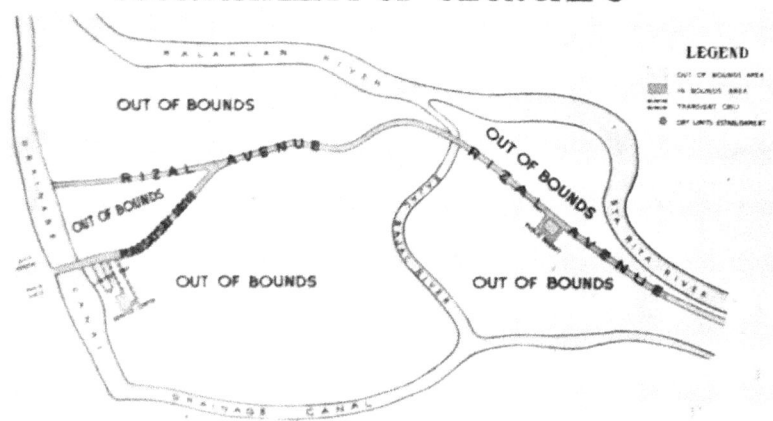

Map identifying to U.S. Sailors and Marines, off limit areas in Olongapo.
USS *Klondike* (AR-22) Western Pacific 1968-1969 cruise book

The ubiquitous jukebox blares away inside a nightclub.
USS *Castor* (AKS-1) 1967 cruise book

Olongapo nightclub singer.

USS *Sanctuary* (AH-17) 1967 cruise book

Zig-zag road to Baguio.
USS *Tappahannock* (AO-43) Western Pacific 1966-1967 cruise book

The routine of shipboard life seemed a million miles away during a Pagsanjan tour. USS *Castor* (AKS-1) 1967 cruise book

A fishing village and indigenous craft that provides a living from the sea.
USS *Arlington* (AGMR-2) 1968 cruise book

Rural areas offer older ways and a slower pace than that found in bustling cities.
USS *Arlington* (AGMR-2) 1968 cruise book

AUSTRALIA

Sydney Harbour Bridge on the starboard bow of USS *Pine Island*, with Fort Denison under it in the background, which is off the entrance to the Australian Naval Base at Garden Island. On the port bow is the Sydney Opera House under construction.
USS *Pine Island* (AV-12) Far East cruise book 1965-1966

Fort Denison, a fortress island named after the then governor and completed in 1857. It was one of the primary defences on Sydney Harbour whose threats then were seen as Russia and France, but was never used for its intended purpose. It is still used today to measure tidal activity.
USS *Arlington* (AGMR-2) 1968 cruise book

Another view of the Opera House and Sydney Harbour Bridge.
USS *Pine Island* (AV-12) Far East 1965-1966 cruise book

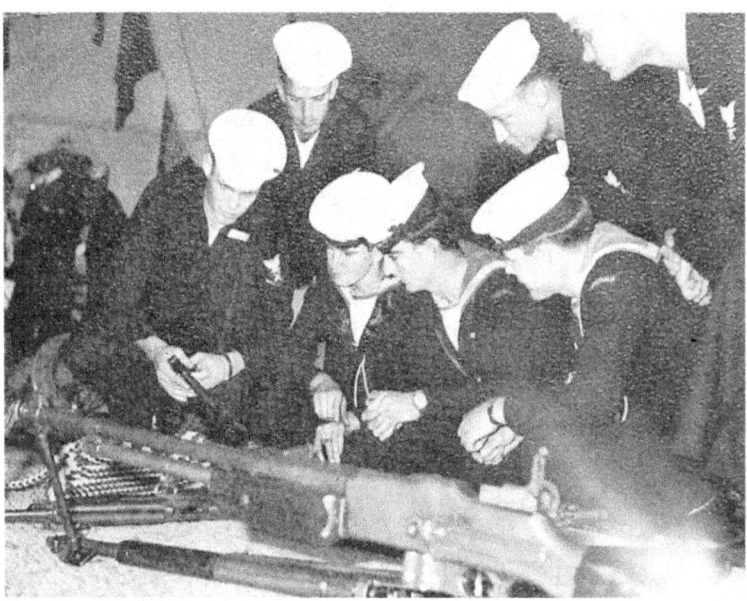

Pine Island sailors explaining USN small arms to their Australian counterparts.
USS *Pine Island* (AV-12) Far East 1965-1966 cruise book

St Mary's Cathedral near Hyde Park, Sydney. Its construction commenced in 1865, it was dedicated in 1882, and finally completed in 1928.
USS *Arlington* (AGMR-2) 1968 cruise book

Excellent view of Sydney Harbour Bridge.
USS *Arlington* (AGMR-2) 1968 cruise book

Australian girls in Sydney were of great interest to many young American sailors. USS *Arlington* (AGMR-2) 1968 cruise book

USS *Arlington* (AGMR-2) 1968 cruise book

HAWAII

Waikiki Beach.
USS *Arlington* (AGMR-2) 1968 cruise book

The beach scene in Hawaii signaled the end of a cruise for the crews of ships home ported at Pearl Harbor, and a welcome interlude for those returning to bases on the West or East Coasts of the United States.
USS *Arlington* (AGMR-2) 1968 cruise book

Appendix A: Justification for MUCs Awarded USS *Cohoes*

CHIEF OF NAVAL OPERATIONS

The Secretary of the Navy takes pleasure in presenting the Meritorious Unit Commendation to

USS COHOES (ANL 78)

For service as set forth in the following

CITATION:

For meritorious service during the periods 4 July through 17 August 1968, 26 September through 20 December 1968, 26 January through 30 April 1969, 11 June through 31 July 1969, 21 August through 8 October 1969, 7 November through 18 December 1969 and from 31 January through 19 March 1970, while assigned to the United States Naval Support Activity, DaNang, Republic of Vietnam. During these periods, USS COHOES, tirelessly and without exception, met all her commitments in the conduct of salvage operations and the repair of petroleum seaload lines throughout the I Corps Tactical Zone in support of Free World Military Assistance Forces. While under heavy enemy artillery attack, COHOES made an extremely difficult but successful salvage of a patrol craft sunk in the Cua Viet Channel. COHOES' speedy clearance of the channel was of great assistance in the movement of vitally needed cargo. In addition, she successfully salvaged two barges and towed free a lighter which had run aground, all within an eight-day period. The ship's superior efforts in repairing a petroleum seaload line in Tan My resulted in a tremendous dollar savings, and prevented several months' loss of the line for replacement. COHOES was successful in salvaging a tug from the Tan My channel. Her success in this venture was particularly noteworthy because several better-equipped salvage vessels were totally unsuccessful. She salvaged over 750 drums of petroleum products from a sunken Vietnamese merchant ship in DaNang Harbor, conducted a survey of YOG 76 which had been mined and sunk in Cua Viet channel, and cleared a mined LCM(8) from Sa Huynh channel. Further, COHOES assisted Naval Support Activity Detachment Cua Viet/Dong Ha in recovery from the effects of Typhoon Doris which had left almost all waterborne assets of the detachment beached. COHOES successfully removed three large barges and several LCPLs and small craft from the beach. She recovered numerous buoys, replaced the POL mooring buoy and repaired the POL seaload line. Her expert performance was a key factor in the quick recovery from the Typhoon by NSAD Cua Viet/Dong Ha. The overall enthusiasm to accomplish each task and the repeated outstanding achievements of the officers and men of USS COHOES attest to their high morale and professionalism, and reflect great credit upon themselves, their ship, and the United States Naval Service.

All personnel attached to and serving on board USS COHOES (ANL-78) during the above-designated periods, or any part thereof, are hereby authorized to wear the Meritorious Unit Commendation Ribbon.

Appendix B: U.S. Seventh Fleet Service Force Ships' Unit Awards

Ship	Combat Action Ribbons	Navy and Meritorious Unit Commendations
Abnaki (ATF-96)		MUC: 1 Sep-2 Oct 69
		MUC: 13-27 Apr 75
Ajax (AR-6)		MUC: 1 Jul 68-1 Oct 69
		NUC: 2-29 Oct 1969
Aludra (AF-55)		
Annapolis (AGMR-1)		MUC: 9 Jan 67-29 Jan 68
Apache (ATF-67)		MUC: 1 Sep 71-23 May 72
		NUC: 3 Feb-7 Oct 69
Arikara (ATF-98)		
Arlington (AGMR-2)		MUC: 11-29 Dec 68
		MUC: 4-31 May 69
		MUC: 20-25 Jul 69
Ashtabula (AO-51)		MUC: 21 May-7 Dec 69
Bellatrix (AF-62)		
Bolster (ARS-38)	7 Feb 68	
	16 Feb 68	
	21 Feb 68	
	4-5 May 70	
Brule (AKL-28)	28 Jan 67	NUC: 16 Mar 66-1 Jan 69
	24 Aug 68	NUC: 3 May-30 Jun 71
Butte (AE-27)		
Cacapon (AO-52)		MUC: 8 Oct 69-8 Apr 70
Caliente (AO-53)		MUC: 24 Jan-11 Aug 72
Camden (AOE-2)		MUC: 26 Jun 68-22 Feb 69
		MUC: 6 Sep 70-21 Mar 71
		NUC: 14 Feb-7 Sep 72
Castor (AKS-1)		
Chanticleer (ASR-7)		
Chara (AE-31)		NUC: 24 Apr-27 Nov 71
Chemung (AO-30)		
Chipola (AO-63)		MUC: 19 May-21 Nov 67
		MUC: 30 May-16 Sep 68
		MUC: 16 Dec 68
		MUC: 17 May-1 Jun 69
		MUC: 23 Sep 70-23 Mar 71
Chowanoc (ATF-100)		
Cimarron (AO-22)		
Cocopa (ATF-101)		
Cohoes (ANL-78)	3 Aug 68	MUC: 4 Jul-17 Aug 68
		MUC: 26 Sep-20 Dec 68
		MUC: 26 Jan-30 Apr 69

		MUC: 11 Jun-31 Jul 69
		MUC: 21 Aug-8 Oct 69
		MUC: 7 Nov-18 Dec 69
		MUC: 31 Jan-19 Mar 70
		MUC: 29 Sep-15 Nov 70
		MUC: 1 May 71-1 Apr 72
		NUC: 1 Jul 70-30 Jun 71
Conserver (ARS-39)		MUC: 30 Nov 70-4 Jan 71
Coucal (ASR-8)		MUC: 4-18 Mar 70
Current (ARS-22)		MUC: 27 Jan-20 Jul 67
Currituck (AV-7)		
Deliver (ARS-23)	27-28 Feb 68	MUC: 13-27 Apr 75
		MUC: 1-7 May 75
Delta (AR-9)		
Diamond Head (AE-19)		MUC: 16-23 Oct 70
Elkhorn (AOG-7)		
Firedrake (AE-14)		
Flint (AE-32)		
Florikan (ASR-9)		
Genesee (AOG-8)	22 Apr 68	NUC: 23 May-25 Sep 65
Graffias (AF-29)		
Grapple (ARS-7)		
Grasp (ARS-24)		MUC: 5 Jul-1 Sep 68
		MUC: 16 Feb-7 Apr 69
		MUC: 4 Jan 70-29 Mar 72
Great Sitkin (AE-17)		
Greenlet (ASR-10)		MUC: 11 Oct 69-14 Apr 70
Guadalupe (AO-32)		MUC: 14 Oct 71-3 Aug 72
Haleakala (AE-25)		MUC: 10 Nov 71-25 Aug 72
Hassayampa (AO-145)		MUC: 20-22 Jul 69
Hector (AR-7)		MUC: 26 Jan-25 Aug 72
Hitchiti (ATF-103)		MUC: 2 Sep-15 Nov 69
Isle Royale (AD-29)		
Jamestown (AGTR-3)		MUC: 1 Nov 65-30 Jun 69
Jason (AR-8)		NUC: 29 Dec 69-10 Jan 70
Kansas City (AOR-3)		MUC: 25 Apr-12 Dec 72
Kawishiwi (AO-146)		MUC: 19 Apr-8 May 75
Kennebec (AO-36)		
Kilauea (AE-26)		MUC: 20 Aug 71-29 Jun 72
Kishwaukee (AOG-9)	21 Feb 68	
Klondike (AR-22)		
Lipan (ATF-85)		MUC: 2 Oct-5 Nov 69
Mahopac (ATA-196)		
Manatee (AO-58)		MUC: 22 Aug 69-9 Feb 70
		MUC: 27 May 72-15 Mar 73
Marias (AO-57)		MUC: 5 May-1 Sep 70
Mark (AKL-12)	15 Feb 67	NUC: 16 Mar 66-1 Jan 69
	19 Nov 69	NUC: 1-31 Jul 69
		NUC: 16 Oct 70-30 Jun 71
Markab (AR-23)		

Appendix B   269

Mars (AFS-1)	17 Sep 66	MUC: 15 Sep 68-28 Aug 69
		MUC: 1 Dec 70-2 Jun 71
		MUC: 1 May-11 Dec 72
		MUC: 12 Apr-11 May 75
Mataco (ATF-86)		
Mattaponi (AO-41)		MUC: 19 Oct 68-15 May 69
Mauna Kea (AE-22)		MUC: 2 Mar-6 Oct 68
		MUC: 10 Jun 70-29 Jun 71
Mauna Loa (AE-8)		
Maury (AGS-16)	11 Dec 65	MUC: 14 Jan-19 Sep 67
Mazama (AE-9)		
Milwaukee (AOR-2)		
Mispillion (AO-105)		MUC: 14 Apr-16 Nov 67
		MUC: 23 Aug 69-28 Feb 70
		MUC: 27 Feb-18 Nov 72
Moctobi (ATF-105)		
Molala (ATF-106)		
Mount Baker (AE-4)		
Mount Hood (AE-29)		MUC: 7 May 72-30 Mar 73
Mount Katmai (AE-16)		MUC: 16 Nov 69-19 Jun 70
		MUC: 24 Apr 72-23 Feb 73
Munsee (ATF-107)		
Navasota (AO-106)		
Neches (AO-47)		
Niagara Falls (AFS-3)		MUC: 24 Jun 72-5 Apr 73
Nitro (AE-23)		MUC: 18 May 72-17 Feb 73
Noxubee (AOG-56)	28 Oct 68	MUC: 10 May-23 Nov 68
	9 Sep 69	
Opportune (ARS-41)		
Oxford (AGTR-1)		MUC: 1 Nov 65-30 Jun 69
Paricutin (AE-18)		
Passumpsic (AO-107)		MUC: 27 Nov 67-23 Jun 68
		MUC: 5 May 72-27 Feb 73
Patapsco (AOG-1)	16 Feb 68	MUC: 1 Oct-31 Dec 67
	27-28 Feb 68	MUC: 26 Jan-31 Mar 68
Pictor (AF-54)		
Piedmont (AD-17)		
Pine Island (AV-12)		
Platte (AO-24)		MUC: 4 Nov 69-21 Apr 70
Pollux (AKS-4)		MUC: 1 Jul 65-1 Jul 67
Ponchatoula (AO-148)		MUC: 8-22 Oct 68
Procyon (AF-61)		
Pyro (AE-24)		MUC: 6 Nov 69-15 May 70
		NUC: 5 Dec 64-23 Oct 65
		NUC: 21 Jan-12 Nov 72
Quapaw (ATF-110)		MUC: 13-27 Apr 75
Rainier (AE-5)		MUC: 11 Jul 68-18 Feb 69
Reclaimer (ARS-42)		MUC: 5 Feb-10 Aug 72
Regulus (AF-57)		

Appendix B

Ship		Award
Rehoboth (AGS-50)		
Repose (AH-16)		NUC: 22 Feb 66-8 Feb 67
		NUC: 9 Feb 67-1 Apr 69
Sacramento (AOE-1)		MUC: 15 Mar-18 Aug 71
Safeguard (ARS-25)		MUC: 17-18 Mar 73
Salisbury Sound (AV-13)		
Samuel Gompers (AD-37)		
San Jose (AFS-7)		MUC: 5 Sep 72-14 Jun 73
Sanctuary (AH-17)		MUC: 11 Apr 69-14 Apr 71
		NUC: 10 Apr 67-10 Apr 69
Santa Barbara (AE-28)		MUC: 29 Jun 72-31 Jan 73
Savannah (AOR-4)		MUC: 15 May-15 Nov 72
Serrano (AGS-24)	3-4 Mar 67	MUC: 14 Jan-19 Sep 67
Shakori (ATF-162)		
Shasta (AE-6)		MUC: 3 Nov 66-16 Apr 67
Sheldrake (AGS-19)		MUC: 19 Oct 67-4 Mar 68
Sioux (ATF-75)		
Sunnadin (ATA-197)		
Suribachi (AE-21)		MUC: 5 Jul 72-27 Feb 73
Taluga (AO-62)		
Tanner (AGS-15)		MUC: 19 Oct 67-4 Apr 68
Tappahannock (AO-43)		
Tawakoni (ATF-114)		MUC: 16 Aug-9 Sep 71
Tawasa (ATF-92)		MUC: 16 Aug-9 Oct 71
Tillamook (ATA-192)		MUC: 1 Dec 66-1 Nov 68
Tolovana (AO-64)		MUC: 29 Sep 72-16 May 73
Tombigbee (AOG-11)		
Towhee (AGS-28)		MUC: 19 Oct 67-4 Mar 68
Ute (ATF-76)		MUC: 27 Aug-15 Nov 69
Vega (AF-59)		MUC: 15 Jul-30 Nov 67
		MUC: 18 Mar-23 Aug 68
		MUC: 22 Apr-7 May 75
Vesuvius (AE-15)		MUC: 4 Oct 69-7 May 70
		NUC: 28 Feb 72-19 Feb 73
Virgo (AE-30)		
Wabash (AOR-5)		
Waccamaw (AO-109)		MUC: 23 Sep-2 Oct 70
		MUC: 27 Jun 72-17 Jan 73
Wandank (ATA-204)		
White Plains (AFS-4)		MUC: 1 Dec 71-12 Jul 72
		MUC: 11 Apr-11 May 75
Wichita (AOR-1)		MUC: 4 Jul 70-19 Jan 71
		MUC: 29 Jul 72-5 Mar 73
Wrangell (AE-12)		
Zelima (AF-49)[1]		

Appendix C: HCU-1 Navy Unit Commendations

THE SECRETARY OF THE NAVY
WASHINGTON

The Secretary of the Navy takes pleasure in commending

HARBOR CLEARANCE UNIT ONE

for service as set forth in the following

CITATION:

For exceptionally meritorious service from 24 February 1966 to 15 March 1967 during combat salvage operations in support of military operations in the Republic of Vietnam. During this period, Harbor Clearance Unit ONE executed over twenty-four hazardous diving and salvage operations, resulting in major contributions to United States efforts in Vietnam. In the face of hostile fire and major obstacles, including heavy seas, strong tidal currents and zero visibility environments during diving operations, unit personnel expeditiously and efficiently accomplished salvage; harbor and river clearance of damaged vessels of all sizes; underwater tasks of all types, including searches for suspected limpet mines; and recovery of aircraft wreckage and enemy ordnance. The outstanding professional competence, ingenuity and personal efforts of the officers and men of Harbor Clearance Unit ONE were in keeping with the highest traditions of the United States Naval Service.

All personnel attached to and serving with Harbor Clearance Unit ONE during the period designated above, or any part thereof, are hereby authorized to wear the Navy Unit Commendation Ribbon.

Secretary of the Navy

THE SECRETARY OF THE NAVY
WASHINGTON

The Secretary of the Navy takes pleasure in commending

HARBOR CLEARANCE UNIT ONE

for service as set forth in the following

CITATION:

For exceptionally meritorious service from 15 March 1967 to 1 July 1969, during salvage operations in support of military operations in the Republic of Vietnam. During this period, Harbor Clearance Unit ONE executed over one hundred and sixteen hazardous diving and salvage operations, resulting in major contributions to United States efforts in the Republic of Vietnam. In the face of hostile fire, sapper attacks, and other major obstacles, including heavy seas, strong tidal current and zero visibility environment during diving operations, Unit personnel expeditiously and efficiently accomplished salvage operations; harbor and river clearance of damaged vessels of all sizes; and underwater tasks of all types, including assistance in removing limpet mines, demolition of obstacles and recovery of aircraft wreckage. By their courage, outstanding professional competence, ingenuity and personal efforts throughout this period, the officers and men of Harbor Clearance Unit ONE upheld the highest traditions of the United States Naval Service.

All personnel attached to and serving with Harbor Clearance Unit ONE during the period designated above, or any part thereof, are hereby authorized to wear the Navy Unit Commendation Ribbon.

John H. Chafee
Secretary of the Navy

Appendix D: ABCD Anthony L. Ey, RAN Letter of Commendation

The Commanding Officer, U. S. Naval Support Facility, DaNang Republic of Vietnam takes pleasure in commending

A. L. EY
ABLE BODIED SEAMAN
ROYAL AUSTRALIAN NAVY

for service as set forth in the following

CITATION

"For meritorious service while attached to the Explosive Ordnance Disposal Team THIRTY-FIVE, and serving with the United States Naval Support Facility, DaNang while conducting intensive salvage operations during the period 3 November to 20 November 1970. Your exemplary professionalism as a member of the on-scene rescue and recovery diving team in salvage operations on the capsized and grounded United States Army Craft YFU 63 near Tan My, in Military Region 1, Republic of Vietnam was outstanding. Your conscientious efforts, endurance and cooperative attitude contributed significantly toward the accomplishment of a very hazardous task. During this period you participated untiringly in a rescue effort which required exceptional courage and professional skill. The personal initiative, sustained outstanding performance and unswerving devotion to duty under exceptionally arduous living and working conditions and the constant threat of enemy attack were in keeping with the highest traditions of the United States Naval Service."

M. A. HORN
Captain, U. S. Navy

Appendix E: USS *Sanctuary* Meritorious Unit Commendation

CHIEF OF NAVAL OPERATIONS

The Secretary of the Navy takes pleasure in presenting the MERITORIOUS UNIT COMMENDATION to

USS SANCTUARY (AH-17)

for service as set forth in the following

CITATION:

For meritorious service from 11 April 1969 to 14 April 1971 while providing highly responsive hospital-ship services to the III Marine Amphibious Force in waters adjacent to the I Corps Tactical Zone of the Republic of Vietnam. During two years of direct support of combat forces in the field, USS SANCTUARY recorded 10,701 helicopter landings on her flight deck; performed over 4,629 major surgical operations; admitted 13,500 patients and treated a total of 35,005 servicemen. More than sixty percent of all patients admitted were subsequently returned to duty due to the excellent medical treatment received. In addition to combat medical support for United States Forces, SANCTUARY provided treatment for Free-World Forces and Vietnamese civilians in urgent need of medical service and humanitarian care. Her record of meritorious service contributed in great measure to United States efforts in Southeast Asia where she was an invaluable sustaining element in direct support of combatant forces. The resourceful professionalism and inspiring devotion to duty demonstrated by the officers and men of the USS SANCTUARY during this period were in keeping with the highest traditions of the United States Naval Service.

For the Secretary,

L. R. Zumwalt, Jr.
Admiral, United States Navy
Chief of Naval Operations

Appendix F: USS *Ajax* and USS *Jason* Navy Unit Commendation

The Secretary of the Navy takes pleasure in presenting the
NAVY UNIT COMMENDATION to

UNITED STATES NAVAL SUPPORT ACTIVITY, SAIGON

for service as set forth in the following

CITATION:
For exceptionally meritorious service from 16 March 1966 to 30 June 1971, in providing logistic support to U.S. Navy, U.S. Coast Guard, and Free World Naval Forces in Military Regions ONE, TWO, THREE and FOUR, Republic of Vietnam. During this period, United States Naval Support Activity, Saigon was tasked with requirements to overhaul and repair craft and equipment; provide administrative personnel, accounting, health, and welfare services; and provide logistical and fiscal support in matters of weapons, ammunitions, base defense, port services, air and surface transportation, and operational publications. Concurrently, programs were initiated to build Riverine Warfare (GAME WARDEN) support bases, develop and expand the Coastal Surveillance (MARKET TIME) and Harbor Defense support bases, develop support for the Mobile Riverine Force concept, and establish policies and procedures for the full range of logistic support. Despite an adverse and often hostile environment, and with a continual shortage of personnel and equipment, United States Naval Support Activity, Saigon provided these necessary support functions to the U.S. Navy and Coast Guard forces operating in all areas of Vietnam. The superlative performance of duty by the personnel of United States Naval Support Activity, Saigon reflects great credit upon themselves and the United States Naval Service.

John W. Warner
Secretary of the Navy

Eligible ships included repair ships USS *Ajax* (2 - 29 October 1969) and USS *Jason* (29 December 1969 - 10 January 1970).

Appendix G: USS *Chara* Meritorious Unit Commendation

CHIEF OF NAVAL OPERATIONS

The Secretary of the Navy takes pleasure in presenting the MERITORIOUS UNIT COMMENDATION to

USS CHARA (AE-31)

for service as set forth in the following

CITATION:

For meritorious service from 24 April 1971 to 27 November 1971, in direct support of United States SEVENTH Fleet combat operations in Southeast Asia. USS CHARA distinguished herself by providing outstanding mobile logistic support to naval units engaged in combat operations against enemy forces in the Republic of Vietnam, and contributed materially to the success of these operations by transferring over 9,300 tons of ammunition to destroyer, cruiser, attack carrier, and other type units during 106 underway replenishments. Through their continuous display of professionalism, determination, resourcefulness, and sheer aggressiveness, the officers and men of USS CHARA contributed immeasurably to the United States mission in Southeast Asia, thereby upholding the highest traditions of the United States Naval Service.

For the Secretary,

E. R. Zumwalt, Jr.
Admiral, United States Navy
Chief of Naval Operations

Appendix H: USS *Mount Katmai* Meritorious Unit Commendation

The Secretary of the Navy takes pleasure in presenting the
MERITORIOUS UNIT COMMENDATION to
USS *MOUNT KATMAI* (AE-16)

for service as set forth in the following

CITATION:
For meritorious service from 16 November 1969 to 19 June 1970 in direct support of United States SEVENTH Fleet combat operations in Southeast Asia. USS *MOUNT KATMAI* distinguished herself by providing outstanding mobile logistic support to naval units engaged in combat operations against enemy forces in the Republic of Vietnam, contributing materially to the success of these operations by transferring over 11,000 tons of ammunition to destroyer, cruiser, and attack carrier units during seventy-six underway replenishments. Through their continuous display of professionalism, determination, and resourcefulness, the officers and men of USS *MOUNT KATMAI* contributed immeasurably to the successful accomplishment of the United States mission in Southeast Asia. Their perseverance and unfailing devotion to duty were in keeping with the highest traditions of the United States Naval Service.

For the Secretary,

E. R. Zumwalt, Jr.
Admiral, United States Navy
Chief of Naval Operations

Appendix I: U.S. 7th Fleet Service Force Personnel Casualties

Date	Ship	Name	Location
5 Feb 66	*Navasota* (AO-106)	SFC Bernard J. Sparenberg	MIA South China Sea – SH3A helicopter passenger
5 Feb 66	*Navasota* (AO-106)	SF1 Glenn E. Asmussen	Same as above
5 Feb 66	*Navasota* (AO-106)	SFM2 Dan D. McConnaugehay	Same as above
17 Oct 66	*Salisbury Sound* (AV-17)	ADR1 Edward G. Salonish	South China Sea
11 Jun 67	*Castor* (AKS-1)	HMCS Frank R. Frost III	South China Sea
4 Jan 68	*Mauna Loa* (AE-8)	Lt (jg) Dennis E. Montague	Gulf of Tonkin
6 Feb 68	*Munsee* (ATF-107)	HM1 John W. Doby	Gulf of Tonkin
16 Mar 68	*Mark* (AKL-12)	BM3 James B. Rickels	Gia Dinh
23 Apr 68	*Genesee* (AOG-8)	SFFN Arthur W. Ball	South China Sea
1 May 68	*Genesee* (AOG-8)	SA Donald R. Schafer	South China Sea – Quang Tri
3 Jun 68	*Vega* (AF-59)	FN James E. Bell	Gulf of Tonkin
16 Dec 68	*Sanctuary* (AH-17)	SN John D. Stuart Jr.	South China Sea
21 Apr 69	*Caliente* (AO-53)	SN Kenneth C. Jenkins	South China Sea
17 Mar 71	*Cohoes* (AN-78)	RMSA Lewis A. Davis	Quang Tin-Chu Lai
15 Jun 71	*Brule* (AKL-28)	FN James A. Souther	Gia Dinh
5 Jun 72	*Ashtabula* (AO-51)	FN David S. Kraner	MIA, Gulf of Tonkin
5 Jun 72	*Chipola* (AO-63)	SN Kylis T. Payne	MIA South China Sea[2]

Bibliography/Chapter Notes

Allan, Donald, John Kennett, *All in the Line of Duty: Honours and Awards to the Royal Australian Navy's Clearance Diving Branch 1951-2018*. Ridgehaven, South Australia: Royal Australian Navy Clearance Divers Association, 2019.

Bartholomew, Charles A., William I. Milwee Jr., *Mud, Muscle, and Miracles: Marine Salvage in the United States Navy*. Washington, DC: Naval History and Heritage Command, Naval Sea Systems Command, 2009.

Bruhn, David D. *Gators Offshore and Upriver: The U.S. Navy's Amphibious Ships and Underwater Demolition Teams, and Royal Australian Navy Clearance Divers in Vietnam*, Berwyn Heights, Md: Heritage Books, ____.

—*Ingram's Fourth Fleet: U.S. and Royal Navy Operations Against German Runners, Raiders, and Submarines in the South Atlantic in World War II*. Berwyn Heights, Md: Heritage Books, 2017.

—*MacArthur and Halsey's "Pacific Island Hoppers": The Forgotten Fleet of World War II*. Berwyn Heights, Md: Heritage Books, 2014.

—*On the Gunline: U.S. Navy and Royal Australian Navy Warships off Vietnam, 1965-1973*. Berwyn Heights, Md: Heritage Books, 2019.

—*Wooden Ships and Iron Men, Volume III: The U.S. Navy's Coastal and Inshore Minesweepers, and the Minecraft That Served in Vietnam, 1953-1976*. Westminster, Md: Heritage Books, 2011.

—*Wooden Ships and Iron Men: The U.S. Navy's Ocean Minesweepers, 1953-1994*. Westminster, Md: Heritage Books, 2006.

Marolda, Edward J. *By Sea, Air, and Land: An Illustrated History of the U.S. Navy and the War in Southeast Asia*. Washington, DC: Naval Historical Center, 1994.

—*The Approaching Storm: Conflict in Asia, 1945-1965*. Washington, DC: Naval History and Heritage Command, 2009

Marolda, Edward J., R. Blake Dunnavent. *Combat at Close Quarters Warfare on the Rivers and Canals of Vietnam*. Washington, DC: Naval History and Heritage Command, 2015.

Mercogliano, Salvatore R. *Fourth Arm of Defense: Sealift and Maritime Logistics in the Vietnam War*. Washington, DC: Naval History and Heritage Command, 2017.

Nott, Rodney, Noel Payne, *The Vung Tau Ferry HMAS Sydney and Escort Ships Vietnam 1965-1972*. Kenthurst NSW, Australia: Rosenberg Publishing, 2008.

Perryman, John, Brett Mitchell, *Australian Navy in Vietnam: Royal Australian Navy Operations 1965-1972*. Silverwater, NSW: Topmill Pty Ltd., 2007.

Sherwood, Darrell, *Nixon's Trident: Naval Power in Southeast Asia, 1968–1972*. Washington, DC: Naval History and Heritage Command, 2009.

Shulimson, Jack, Leonard A. Blasiol, Charles R. Smith, David A. Dawson, *U.S. Marines in Vietnam: The Defining Year 1968*. Washington, DC: History and Museums Division, Headquarters U.S. Marine Corps, 1997.

Wildenberg, Thomas, *Gray Steel and Black Oil Fast Tankers and Replenishment at Sea in the U.S. Navy, 1912-1995*. Annapolis, Md: Naval Institute, 1996.

PREFACE NOTES:

[1] Oral History Interview of Capt. Leon Grabowsky, USN (Retired), June 7, 1991, by Donald R. Lennon, East Carolina University Manuscript Collection (https://digital.lib.ecu.edu/text/11270: accessed 9 August 2019).

[2] "Carriers: Airpower at Sea" by Arnold E. van Beverhoudt Jr. (https://www.sandcastlevi.com/sea/carriers/intro.html: accessed 6 August 2019)

[3] Ibid.

[4] Interview of Capt. Leon Grabowsky.

[5] Salvatore R. Mercogliano, *Fourth Arm of Defense: Sealift and Maritime Logistics in the Vietnam War* (Washington, DC: Naval History and Heritage Command, 2017), 13, 31.

[6] Interview of Capt. Leon Grabowsky; Mercogliano, *Fourth Arm of Defense*, 26.

[7] "Life Aboard an AOG" by Paul Gryniewicz (http://www.ussnoxubee.org/paulgryn2.htm: accessed 15 July 2019).

[8] Charles A. Bartholomew, and William I. Milwee Jr., *Mud, Muscle, and Miracles: Marine Salvage in the United States Navy* (Washington, DC: Naval History and Heritage Command, Naval Sea Systems Command, 2009), 258, 263, 266.

[9] Ibid.

[10] Commander, U.S. Naval Forces Vietnam, Monthly Historical Supplement, August 1968, 4 September 1968.

[11] *Bolster* deck logs for February 1968, and May 1970.

[12] Edward J. Marolda, *By Sea, Air, and Land*, Chapter 3: The Years of Combat, 1965-1968 (https://www.history.navy.mil/research/library/online-reading-room/title-list-alphabetically/b/by-sea-air-land-marolda/chapter-3-the-years-of-combat-1965-1968.html: accessed 9 August 2019).

[13] Ibid.

[14] "Naval Operations in Vietnam" by Jozef Straczek (http://www.navy.gov.au/history/feature-histories/naval-operations-vietnam: accessed 19 July 2019).

CHAPTER 1 NOTES:
[1] USS *Brule* and USS *Garrett County* deck logs for 28 August 1968; phone conversation with Thomas Young of 14 August 2019.
[2] USS *Garrett County* deck log for 28 August 1968.
[3] David D. Bruhn, *MacArthur and Halsey's "Pacific Island Hoppers": The Forgotten Fleet of World War II* (Berwyn Heights, Md: Heritage Books, 2014), xxiii-xxiv, xxvii, 259-260.
[4] Ibid, xxvii.
[5] Ibid.
[6] Former *Brule* crewmember Dick Moody (https://www.mrfa.org/us-navy/us-navy-mobile-riverine-force/brule-akl-28/: accessed 10 August 2019).
[7] Ibid.
[8] USS *Brule* deck log for 28 August 1968; phone conversation with Thomas Young of 14 August 2019.
[9] Phone conversation with Thomas Young of 14 August 2019.
[10] USS *Brule* deck log for 28 August 1968.
[11] Phone conversation with Thomas Young of 14 August 2019.
[12] USS *Brule* deck log for 28 August 1968.

CHAPTER 2 NOTES:
[1] David D. Bruhn, *Wooden Ships and Iron Men, Volume III: The U.S. Navy's Coastal and Inshore Minesweepers, and the Minecraft That Served in Vietnam, 1953-1976* (Westminster, Md: Heritage Books, 2011), 111.
[2] Ibid, 111-112.
[3] Ibid, 112.
[4] Ibid.
[5] Ibid.
[6] Ibid, 113.
[7] Ibid.
[8] Ibid, 113-114.
[9] Ibid. 114.
[10] Ibid.
[11] Ibid, 122-123.
[12] Ibid, 123.

CHAPTER 3 NOTES:
[1] "Lyndon Baines Johnson Inaugural Address, Wednesday, January 20, 1965" (http://www.let.rug.nl/usa/presidents/lyndon-baines-johnson/inaugural-address-1965.php: accessed 29 July 2019).
[2] "The Vietnam War: The Jungle War 1965-1968" (http://www.historyplace.com/unitedstates/vietnam/index-1965.html; "H-

009-3: Significant U.S. Navy Operations and Events in Vietnam Through 1967" (https://www.history.navy.mil/content/history/nhhc/about-us/leadership/director/directors-corner/h-grams/h-gram-009/h009-3.html: both accessed 29 July 2019).
[3] Ut supra.
[4] "H-017-2: Rolling Thunder—A Short Overview" (https://www.history.navy.mil/content/history/nhhc/about-us/leadership/director/directors-corner/h-grams/h-gram-017/h-017-2.html: accessed 29 July 2019).
[5] Edward J. Marolda, *The Approaching Storm: Conflict in Asia, 1945-1965* (Washington, DC: Naval History and Heritage Command, 2009), 76.
[6] Edward J. Marolda and R. Blake Dunnavent, *Combat at Close Quarters Warfare on the Rivers and Canals of Vietnam* (Washington, DC: Naval History and Heritage Command, 2015), 2.
[7] "H-009-3: Significant U.S. Navy Operations and Events in Vietnam Through 1967."
[8] Marolda and Dunnavent, *Combat at Close Quarters Warfare on the Rivers and Canals of Vietnam*, 3.
[9] "H-009-3: Significant U.S. Navy Operations and Events in Vietnam Through 1967."
[10] "H-009-3: Significant U.S. Navy Operations and Events in Vietnam Through 1967"; Mercogliano, *Fourth Arm of Defense*, 12.
[11] "The Vietnam War: The Jungle War 1965-1968"; "H-009-3: Significant U.S. Navy Operations and Events in Vietnam Through 1967."

CHAPTER 4 NOTES:
[1] "Royal Australian Navy, HMAS *Sydney*" (https://anzacportal.dva.gov.au/history/special-features/veterans-stories/vietnam-war-stories/bill-kane: accessed 29 July 2019).
[2] "How eight countries got bogged down in the Vietnam War's Cold War proxy battle" by Jesse Greenspan (https://www.history.com/news/vietnam-war-combatants#section_7: accessed 16 August 2019).
[3] "HMAS *Sydney* (III)" (http://www.navy.gov.au/hmas-sydney-iii: accessed 19 July 2019).
[4] "Vung Tau Ferry" (https://anzacportal.dva.gov.au/history/conflicts/australia-and-vietnam-war/australia-and-vietnam-war/royal-australian-navy/vung-tau: accessed 29 July 2019).
[5] "The *Majestic* Class Light Fleet Aircraft Carrier HMAS *Sydney*" (https://www.awm.gov.au/collection/044798: accessed 19 July 2019).
[6] "The *Majestic* Class Light Fleet Aircraft Carrier HMAS *Sydney*;" "HMAS *Sydney* (III)."
[7] "HMAS *Sydney* (III)."
[8] Ibid.
[9] Ibid.
[10] Ibid.

[11] Ibid.

CHAPTER 5 NOTES:
[1] "Commissioned for 69 Days-HMAS *Boonaroo*" (https://www.navyhistory.org.au/commissioned-for-69-days-hmas-boonaroo/: accessed 19 July 2019).
[2] "HMAS *Boonaroo*" (http://www.navy.gov.au/hmas-boonaroo: accessed 19 July 2019).
[3] "HMAS *Boonaroo*"; "Commissioned for 69 Days-HMAS *Boonaroo*."
[4] Ut supra.
[5] "Commissioned for 69 Days-HMAS *Boonaroo*."
[6] Ibid.
[7] Ibid.
[8] Ibid.
[9] Ibid.
[10] "Naval Operations in Vietnam" by Jozef Straczek (http://www.navy.gov.au/history/feature-histories/naval-operations-vietnam); "HMAS *Jeparit*" (http://www.nepeannaval.org.au/Museum/General-Purpose-Vessels/HMAS-Jeparit.html); "HMAS *Jeparit*" (http://www.navy.gov.au/hmas-jeparit: accessed 20 July 2019).
[11] Ut supra.
[12] David D. Bruhn, *Ingram's Fourth Fleet: U.S. and Royal Navy Operations Against German Runners, Raiders, and Submarines in the South Atlantic in World War II* (Berwyn Heights, Md: Heritage Books, 2017), 59.
[13] Rodney Nott and Noel Payne, *The Vung Tau Ferry HMAS Sydney and Escort Ships Vietnam 1965-1972* (Kenthurst NSW, Australia: Rosenberg Publishing, 2008), 168; Bruhn, *Ingram's Fourth Fleet: U.S. and Royal Navy Operations Against German Runners, Raiders, and Submarines in the South Atlantic in World War II*, 27, 36.
[14] Nott and Payne, *The Vung Tau Ferry HMAS Sydney and Escort Ships Vietnam 1965-1972*, 168; Bruhn, *Ingram's Fourth Fleet*, 62-71.
[15] Bruhn, *Ingram's Fourth Fleet: U.S. and Royal Navy Operations Against German Runners, Raiders, and Submarines in the South Atlantic in World War II*, 62-71.
[16] Ibid.
[17] Nott and Payne, *The Vung Tau Ferry HMAS Sydney and Escort Ships Vietnam 1965-1972*, 168.

CHAPTER 6 NOTES:
[1] USS *Maury* (AGS-16) and USS *Serrano* (AGS-24) Vietnam Survey 1965-66 cruise book.
[2] "Our Heritage New Look at Vietnam War Support" by Martin K. Gordon (https://www.nga.mil/About/History/NGAinHistory/Documents/.../Vietnam.pdf: accessed 11 July 2019).
[3] USS *Maury* (AGS-16) and USS *Serrano* (AGS-24) Vietnam Survey 1965-66 cruise book.

[4] Ibid.
[5] Untitled, declassified document (https://www.vietnam.ttu.edu/star/images/1201/1201001040d.pdf: accessed 8 July 2019).
[6] Ibid.
[7] David D. Bruhn, *Wooden Ships and Iron Men: The U.S. Navy's Ocean Minesweepers, 1953-1994* (Westminster, Md: Heritage Books, 2006), 144-145.
[8] USS *Maury* (AGS-16) and USS *Serrano* (AGS-24) Vietnam Survey 1965-66 cruise book; "Apr 27, 1521 CE: Magellan Killed in Philippine Skirmish" (https://www.nationalgeographic.org/thisday/apr27/magellan-killed-philippine-skirmish/: accessed 9 July 2019).
[9] USS *Maury* (AGS-16) and USS *Serrano* (AGS-24) Vietnam Survey 1965-66 cruise book.
[10] Ibid.
[11] Ibid.
[12] Ibid.
[13] Ibid.
[14] Ibid.
[15] Ibid.
[16] Ibid.
[17] USS *Maury* (AGS-16) and USS *Serrano* (AGS-24) Vietnam Survey 1965-66 cruise book; "USS *Maury* AG16 Thanks for the Memories 1945-1969" (http://www.ussmauryags16.org/memories.html: accessed 6 July 2019).
[18] USS *Maury* (AGS-16) and USS *Serrano* (AGS-24) Vietnam Survey 1965-66 cruise book.
[19] Ibid.
[20] Ibid.
[21] Ibid.
[22] Ibid.
[23] Ibid.
[24] Ibid.
[25] Ibid.
[26] *Rehoboth, DANFS*.

CHAPTER 7 NOTES:
[1] "USS *Maury* AG16 Thanks for the Memories 1945-1969."
[2] "A History of the USS *Maury* AGS16" (http://www.ussmauryags16.org/Maury%27s_History.html: accessed 6 July 2019).
[3] USS *Maury* (AGS-16) Kwajalein – Vietnam Survey 1966-1967 cruise book.
[4] Ibid.
[5] USS *Maury* (AGS-16) Kwajalein – Vietnam Survey 1966-1967 cruise book; *Serrano, DANFS*.
[6] USS *Maury* (AGS-16) Kwajalein – Vietnam Survey 1966-1967 cruise book.
[7] "USS *Maury* AG16 Thanks for the Memories 1945-1969."
[8] Ibid.

[9] Ibid.
[10] USS *Maury* (AGS-16) Kwajalein – Vietnam Survey 1966-1967 cruise book.
[11] Ibid.
[12] Ibid.
[13] Ibid.

CHAPTER 8 NOTES:
[1] USS *Tanner* (AGS-15) 1967-68 cruise book.
[2] USS *Tanner* (AGS-15) 1967-68 cruise book; *Towhee, DANFS*.
[3] *Towhee, DANFS*.
[4] USS *Tanner* (AGS-15) 1967-68 cruise book; Marolda, *By Sea, Air, and Land*, Chapter 3: The Years of Combat, 1965-1968.
[5] "Mamie Van Doren Biography" (https://www.imdb.com/name/nm0886638/bio: accessed 13 July 2019).
[6] "A History of the USS *Maury* AGS16."

CHAPTER 9 NOTES:
[1] "Sub Hunts in a Seaplane: On cold war missions, in any weather, Martin Marlins prowled the globe" (https://www.airspacemag.com/military-aviation/cold-war-marlin-pilots-180960970/: accessed 15 August 2019).
[2] USS *Pine Island* (AV-12) Far East cruise book 1965-1966.
[3] Ibid.
[4] "USS *Currituck* AV-7 The Wild Goose Lives On" (https://web.archive.org/web/20070928230257/http://www.usscurrituck.org/); "USS *Salisbury Sound* (AV-13)" (http://www.salisburysound.org/index.php/history/: accessed 12 August 2019: both accessed 12 August 2019); *Pine Island, DANFS*.
[5] "USS *Currituck* AV-7 The Wild Goose Lives On"; "USS *Salisbury Sound* (AV-13)."
[6] USS *Pine Island* (AV-12) Far East 1965-1966 cruise book.
[7] "USS *Salisbury Sound* (AV-13)."
[8] "My time aboard the Sally" by Frank Giesler (http://www.salisburysound.org/index.php/sea_stories/: accessed 12 August 2019).
[9] Ibid.
[10] "Sub Hunts in a Seaplane."
[11] "A Look Back at the Last Days of the Mighty Martin Marlin" (http://warbirdsnews.com/warbirds-news/martin-marlin.html: accessed 15 August 2019).
[12] "Sub Hunts in a Seaplane."
[13] Ibid.
[14] "USS *Salisbury Sound* (AV-13)."
[15] "Sub Hunts in a Seaplane."
[16] "USS *Currituck* AV-7 The Wild Goose Lives On."

CHAPTER 10 NOTES:
[1] "Memories" (http://www.ussjamestown.com/memories1.html: accessed 13 July 2019).
[2] Roland Nino Martinez correspondence of 19 August 2019.
[3] Ibid.
[4] "Technical research ship" (https://military.wikia.org/wiki/Technical_research_ship: accessed 13 July 2019).
[5] *Oxford, Jamestown, DANFS*; "Memories," USS *Jamestown* AGTR-3" (http://www.ussjamestown.com/: accessed 13 Jul 2019).
[6] *Oxford, Jamestown, DANFS*; "Technical Research Ship (AGTR) Index" (http://www.navsource.org/archives/09/60/60idx.htm: accessed 13 July 2019).
[7] "Members of the United States Intelligence Board 1962-63" by David Coleman (https://historyinpieces.com/research/members-united-states-intelligence-board: accessed 13 July 2019).
[8] *Oxford, DANFS*; Chief of Naval Operations, Master List of Unit Awards and Campaign Medals, OPNAV NOTICE 1650, 9 March 2001.

CHAPTER 11 NOTES:
[1] USS *Annapolis* (AGMR-1) Western Pacific 1967 cruise book.
[2] "USS Annapolis - AGMR-1" (http://www.agmr1-uss-annapolis.org/: accessed 23 August 2019).
[3] Ibid.
[4] Ibid.
[5] Ibid.
[6] Ibid.
[7] USS *Annapolis* (AGMR-1) Western Pacific 1967 cruise book; "USS Annapolis - AGMR-1."
[8] "USS Arlington.com" (http://www.ussarlington.com/: accessed 23 August 2019).
[9] "Detailed Report of TF 130 Participation in APOLLO 11 Mission" (https://www.history.navy.mil/content/dam/nhhc/research/archives/apollo-11/tf-130-apollo-11.pdf: accessed 23 August 2019).
[10] *Arlington, DANFS*.
[11] "Navy Recovery Ships" (https://history.nasa.gov/ships.html: accessed 24 August 2019).
[12] *Arlington, DANFS*; "Apollo 10 and NASA — Navy Collaboration in Search and Recovery Operations" (https://www.history.navy.mil/content/history/nhhc/browse-by-topic/exploration-and-innovation/navy-and-space-exploration/Apollo_10_NASA.html: accessed 24 August 2019).
[13] "Apollo 10 and NASA — Navy Collaboration in Search and Recovery Operations."
[14] Ibid.

[15] USS *Arlington* (AGMR-2) Middle Pacific 1969 cruise book; Bruhn, *Wooden Ships and Iron Men: The U.S. Navy's Ocean Minesweepers, 1953-1994*, 111.
[16] *Arlington, DANFS*.
[17] "President Nixon and President Thieu Meet at Midway Island, June 8, 1969" (https://www.nixonfoundation.org/2014/06/president-nixon-president-thieu-meet-midway-island-june-8-1969/: accessed 24 August 2019).
[18] USS *Arlington* (AGMR-2) Middle Pacific 1969 cruise book.
[19] "President Nixon and President Thieu Meet at Midway Island, June 8, 1969"; "Vietnam War Allied Troop Levels 1960-73" (http://www.americanwarlibrary.com/vietnam/vwatl.htm: accessed 24 August 2019).
[20] "What was Nixon's Vietnamization Policy?" (https://thevietnamwar.info/what-was-nixons-vietnamization-policy/: accessed 25 August 2019).
[21] "First U.S. troops withdrawn from South Vietnam" (https://www.history.com/this-day-in-history/first-u-s-troops-withdrawn-from-south-vietnam: accessed 25 August 2019).
[22] "H-033-4: 50th Anniversary of the First Moon Landing" (https://www.history.navy.mil/content/history/nhhc/about-us/leadership/director/directors-corner/h-grams/h-gram-033/h-033-4.html: accessed 23 August 2019).
[23] Ibid.
[24] "H-033-4: 50th Anniversary of the First Moon Landing"; *Arlington, DANFS*.
[25] Ut supra.
[26] "USS *Annapolis* - AGMR-1."
[27] Naval Communications Bulletin OPNAV 94-P2, December 1968.

CHAPTER 12 NOTES:

[1] "A Few More Bits and Pieces" by Rik Kuhnfew (http://www.ussnoxubee.org/memories.htm: accessed 14 July 2019).
[2] "Gasoline Tanker (AOG) Index" (http://www.navsource.org/archives/09/20/20idx.htm: accessed 14 July 2019).
[3] Chief of Naval Operations, Master List of Unit Awards and Campaign Medals, OPNAV NOTICE 1650, 9 March 2001.
[4] "Life Aboard an AOG" by Paul Gryniewicz (http://www.ussnoxubee.org/paulgryn2.htm: accessed 15 July 2019).
[5] USS *Kishwaukee* (AOG-9) Western Pacific 1967-68 cruise book; "I Corps" (https://namwartravel.com/i-corps/: accessed 5 May 2019).
[6] USS *Kishwaukee* (AOG-9) Western Pacific 1967-68 cruise book.
[7] "AOGs- gas pipeline at sea" by Mark Tempest (http://www.eaglespeak.us/2009/02/sunday-ship-history.html: accessed 15 July 2019).
[8] USS *Kishwaukee* (AOG-9) Western Pacific 1967-68 cruise book.
[9] Ibid.

294　Bibliography/Chapter Notes

[10] Ibid.
[11] Ibid.
[12] Ibid.
[13] Ibid.
[14] Naval Support Activity Da Nang, 1966-67 cruise book.
[15] Marolda, *By Sea, Air, and Land*, Chapter 3: The Years of Combat, 1965-1968.
[16] USS *Kishwaukee* (AOG-9) Western Pacific 1967-68 cruise book.
[17] Ibid.
[18] "A Vietnam History" by John Heatherman (http://www.navsource.org/archives/09/20/2008v.htm: accessed 14 July 2019).
[19] "A Vietnam History" by John Heatherman; Jack Shulimson, Leonard A. Blasiol, Charles R. Smith, and David A. Dawson, *U.S. Marine Corps U.S. Marines in Vietnam The Defining Year 1968* (Washington, DC: History and Museums Division, Headquarters U.S. Marine Corps, 1997), 38.
[20] "A Vietnam History" by John Heatherman.
[21] "Arthur Wyman Ball" (https://www.virtualwall.org/db/BallAW01a.htm: accessed 15 July 2019).
[22] "A Vietnam History" by John Heatherman; "Arthur Wyman Ball."
[23] "Incoming!" by Paul Grynieiwcz (http://www.ussnoxubee.org/: accessed 15 July 2019).
[24] Ibid.
[25] Ibid.
[26] Ibid.
[27] "Mine!" by Paul Gryniewicz (http://www.ussnoxubee.org/: accessed 16 July 2019).
[28] Ibid.
[29] "Sapper Attack: The Elite North Vietnamese Units" by Arnold Blumberg, *Vietnam Magazine*, 1 February 2017.
[30] Ibid.
[31] "Mine!" by Paul Gryniewicz.
[32] Ibid.
[33] Ibid.
[34] Ibid.
[35] Ibid.
[36] Ibid.
[37] Ibid.

CHAPTER 13 NOTES:
[1] USS *Castor* (AKS-1) 1967 cruise book; *Castor, DANFS*.
[2] Ut supra.
[3] USS *Castor* (AKS-1) 1967 cruise book.
[4] Ibid.
[5] Ibid.
[6] USS *Pictor* (AF-54) South China Sea 1965 cruise book.

[7] Ibid.
[8] Ibid.
[9] Ibid.
[10] Ibid.
[11] "AFS-1 *Mars* Combat Stores Ship" https://www.globalsecurity.org/military/systems/ship/afs-1.htm: accessed 28 August 2019).
[12] "Combat Stores Ship (AFS) Index" (http://www.navsource.org/archives/09/52/52idx.htm: accessed 28 August 2019).

CHAPTER 14 NOTES:
[1] Bartholomew and Milwee Jr., *Mud, Muscle, and Miracles*, 247.
[2] Ibid, 250-251.
[3] Ibid, 251.
[4] Ibid, 253, 262.
[5] Ibid, 263.
[6] Bartholomew and Milwee Jr., *Mud, Muscle, and Miracles*, 253; "USS *Frank Knox* (DDR-742), 1944-1971 – Grounding and Salvage, July-August 1965" (https://www.ibiblio.org/hyperwar/OnlineLibrary/photos/sh-usn/usnsh-f/dd742-1.htm: accessed 1 September 2019).
[7] Bartholomew and Milwee Jr., *Mud, Muscle, and Miracles*, 253-254.
[8] Ibid, 254.
[9] Ibid.
[10] Ibid.
[11] Ibid, 255.
[12] Ibid.
[13] "USS *Frank Knox* (DDR-742), 1944-1971 – Grounding and Salvage, July-August 1965."
[14] Ibid.
[15] Bartholomew and Milwee Jr., *Mud, Muscle, and Miracles*, 258.
[16] *Terrell County*, *DANFS*.
[17] Bartholomew and Milwee Jr., *Mud, Muscle, and Miracles*, 259; *Terrell County*, *DANFS*.
[18] "United States Naval Operations Vietnam, Highlights; February 1966" (https://www.history.navy.mil/content/history/nhhc/research/archives/digitized-collections/vietnam-war/united-states-naval-operations-vietnam-highlights-february-1966.html: accessed 2 September 2019).
[19] Bartholomew and Milwee Jr., *Mud, Muscle, and Miracles*, 259; "United States Naval Operations Vietnam, Highlights; February 1966"; "Highlights March 1966" (https://www.history.navy.mil/research/archives/digitized-collections/vietnam-war/highlights-march-1966.html#salvage: accessed 2 September 2019).
[20] "Highlights March 1966."

[21] "MS *Amastra*'s perilous duty on behalf of the US at Nha Trang" by Howard E. Bartholf (https://www.historynet.com/ms-amastras-perilous-duty-behalf-us-nha-trang.htm: accessed 2 September 2019).
[22] Ibid.
[23] Ibid.
[24] Ibid.
[25] Ibid.
[26] Ibid.
[27] Commander, U.S. Naval Forces Vietnam, Monthly Historical Supplement June 1967, 17 September 1967.
[28] "U.S. Naval Support Activity Cua Viet (1967-1970)" (https://www.mrfa.org/us-navy/us-navy-mobile-riverine-force/u-s-naval-bases-support-activities-vietnam/cua-viet-u-s-naval-support-activity-1967-1970/: accessed 3 September 2019).
[29] Ibid.
[30] Commander, U.S. Naval Forces Vietnam, Monthly Historical Supplement, June 1967, 17 September 1967; Bartholomew and Milwee Jr., *Mud, Muscle, and Miracles*, 259.
[31] Commander, U.S. Naval Forces Vietnam, Monthly Historical Supplement, October 1967, 19 February 19[68].
[32] Commander, U.S. Naval Forces Vietnam, Monthly Historical Supplement, October 1967, 19 February 19[68]; Bartholomew and Milwee Jr., *Mud, Muscle, and Miracles*, 259.
[33] Bartholomew and Milwee Jr., *Mud, Muscle, and Miracles*, 262.

CHAPTER 15 NOTES:
[1] David D. Bruhn, *On the Gunline: U.S. Navy and Royal Australian Navy Warships off Vietnam, 1965-1973* (Berwyn Heights, Md: Heritage Books, 2019), xv-xvii.
[2] Ibid.
[3] David D. Bruhn, *Gators Offshore and Upriver: The U.S. Navy's Amphibious Ships and Underwater Demolition Teams, and Royal Australian Navy Clearance Divers in Vietnam* (Berwyn Heights, Md: Heritage Books, 20__), xviii, xxxix, xl; Commodore Hector Donohue, AM RAN (Retired) correspondence of 2 October 2019.
[4] Bruhn, *Gators Offshore and Upriver*, xviii.
[5] Don 'Scotty' Allan and John Kennett, *All in the Line of Duty: Honours and Awards to the Royal Australian Navy's Clearance Diving Branch 1951-2018* (Ridgehaven, South Australia: Royal Australian Navy Clearance Divers Association, 2019), 32.
[6] Bruhn, *Wooden Ships and Iron Men, Volume III*, 4.
[7] Ibid, 5.
[8] Ibid.
[9] Allan and Kennett, *All in the Line of Duty*, 32.
[10] Bruhn, *Wooden Ships and Iron Men, Volume III*, 6.
[11] Ibid, 7.
[12] Ibid.

[13] Allan and Kennett, *All in the Line of Duty*, 63; John Perryman and Brett Mitchell, *Australian Navy in Vietnam: Royal Australian Navy Operations 1965-1972* (Silverwater, NSW: Topmill Pty Ltd., 2007), 80-83.
[14] Allan and Kennett, *All in the Line of Duty*, 63; Perryman and Mitchell, *Australian Navy in Vietnam*, 80-83.
[15] U.S. Navy Commendation Medal award recommendation for Lt. Edward Linton, RAN.
[16] Allan and Kennett, *All in the Line of Duty*, 63.
[17] Allan and Kennett, *All in the Line of Duty*, 64; Perryman and Mitchell, *Australian Navy in Vietnam*, 80-83.
[18] U.S. Navy Commendation Medal award recommendation for Lt. Edward Linton, RAN.
[19] Bruhn, *Gators Offshore and Upriver*, 278, 285.
[20] Ibid, 281, 283.

CHAPTER 16 NOTES:
[1] DA, Headquarters, XXIV Corps, APO 96349, Award of the Meritorious Unit Commendation; recommendation for, 19 April 1971.
[2] USS *Repose* (AH-16) Republic of Vietnam 1969-1970 cruise book.
[3] Ibid.
[4] "HMS *Amethyst* Incident, Yangtse River, April to May 1949" (http://www.naval-history.net/WXLG-Amethyst1949.htm: accessed 12 September 2019).
[5] Ibid.
[6] Ibid.
[7] Ibid.
[8] Ibid.
[9] USS *Repose* (AH-16) Republic of Vietnam 1969-1970 cruise book.
[10] Ibid.
[11] "HMS *Amethyst* Incident, Yangtse River, April to May 1949."
[12] USS *Repose* (AH-16) Republic of Vietnam 1969-1970 cruise book.
[13] USS *Sanctuary* (AH-17) 1967 cruise book.
[14] Ibid.
[15] Ibid.
[16] Ibid.
[17] Ibid.
[18] Ibid.
[19] *Sanctuary*, *DANFS*.
[20] USS *Sanctuary* (AH-17) 1967 cruise book.
[21] Ibid.
[22] Ibid.
[23] Ibid.
[24] Ibid.
[25] Ibid.
[26] *Repose*, *DANFS*.
[27] USS *Sanctuary* (AH-17) 1967 cruise book.

[28] Ibid.

CHAPTER 17 NOTES:
[1] USS *Markab* (AR-23) Western Pacific 1967 cruise book.
[2] Chief of Naval Operations, Master List of Unit Awards and Campaign Medals, OPNAVNOTE 1650 N09B1 of 9 March 2001, credits USS *Prairie* (AD-15) with being in Vietnam waters on 16-17 March 1973, but this is not reflected in her 1973 Western Pacific cruise book.
[3] "Ships Docked to Vietnam Shore or Pier" (https://www.vetshq.com/ships-docked-to-vietnam-shore-pier/: accessed 16 September 2019).
[4] Commander, U.S. Naval Forces, Vietnam, Monthly Historical Supplement, January 1967, 20 March 1967.
[5] Commander, U.S. Naval Forces, Vietnam, Monthly Historical Supplement, January 1967, 20 March 1967; *Mahnomen County*, DANFS.
[6] USS *Markab* (AR-23) Western Pacific 1967 cruise book.
[7] *Ajax*, DANFS.
[8] Bruhn, *Wooden Ships and Iron Men, Volume III*, 1-2.
[9] Ibid, 1.
[10] Ibid, 2-3.
[11] Ibid, 3.
[12] Correspondence with Mathew Zimmer of 18 September 2019.
[13] *Ajax*, DANFS.
[14] Ibid. The *Directory of American Naval Fighting Ships* cites *Ajax*'s duty at Vung Tau in 1970 as from 13 April to 9 May, while the Master List of Unit Awards and Campaign Medals, OPNAV NOTICE 1650 of 9 March 2001, credits her with 18 April-11 May 1971.
[15] "Gary James' Interview with John Claude Gummoe of The Cascades" (http://www.classicbands.com/CascadesInterview.html); "John Claude Gummoe" (http://www.rhythmoftherain.com/bio.html); "The Very Best of The Cascades" (http://www.rhythmoftherain.com/bestof.html: all accessed 20 September 2019).
[16] Ut supra.
[17] Ut supra.
[18] "The Very Best of The Cascades"; "Gary James' Interview with John Claude Gummoe of The Cascades"; "Billboard Announces Top 100 Songs of the Century" (https://www.bmi.com/news/entry/232893: accessed 20 September 2019).
[19] USS *Klondike* (AR-22) Western Pacific 1969 cruise book.
[20] USS *Delta* (AR-9) Western Pacific 1969 cruise book; "Ships Docked to Vietnam Shore or Pier" (https://www.vetshq.com/ships-docked-to-vietnam-shore-pier/: accessed 16 September 2019); Don Wenz (http://brownwater-navy.com/guestbook/Gbook1600.htm: accessed 18 September 2019).
[21] "USS *Hector* AR-7 Association" (http://www.usshector.com/index.html: accessed 20 September 2019).
[22] USS *Samuel Gompers* (AD-37) Western Pacific 1971-1972 cruise book.

[23] Citation Nr: 0417356 (https://www.va.gov/vetapp04/files2/0417356.txt: accessed 20 September 2019).

CHAPTER 18 NOTES:
[1] "Class: *Lassen* (AE-3)" (http://www.shipscribe.com/usnaux/AE/AE03.html: accessed 11 October 2019).
[2] "Ships of the U.S. Navy, 1940-1945" (https://www.ibiblio.org/hyperwar/USN/USN-ships.html#aeex); "USS *Mount Baker* (AE-4)" (http://www.navsource.org/archives/09/05/0504.htm: both accessed 9 October 2019).
[3] USS *Mazama* (AE-9) Western Pacific 1966 cruise book.
[4] Ibid.
[5] Ibid.
[6] Ibid.
[7] *Virgo, DANFS*.
[8] USS *Virgo* (AE-30) Western Pacific 1970 cruise book.
[9] *Virgo, DANFS*.
[10] *Chara, DANFS*.
[11] Ibid.
[12] "USS Mount Hood and crew lost in massive explosion" (http://ww2today.com/10-november-1944-the-crew-of-uss-mount-hood-disappear-in-massive-explosion: accessed 9 October 2019).
[13] Ibid.
[14] "Ships of the U.S. Navy, 1940-1945" (https://www.ibiblio.org/hyperwar/USN/USN-ships.html#ae: accessed 9 October 2019)
[15] *Rainier, DANFS*.
[16] Ibid.
[17] USS *Vesuvius* Vietnam 1969-1970 cruise book.
[18] Ibid.
[19] *Suribachi, Mauna Kea, DANFS*.
[20] Decommissioning Ceremony in honor of USS *Nitro* (AE-23) booklet - Earle, New Jersey April 28, 1995.
[21] "William Loren McGonagle 19 November 1925 - 3 March 1999" (https://www.history.navy.mil/content/history/nhhc/research/library/research-guides/modern-biographical-files-ndl/modern-bios-m/mcgonagle-william-loren.html: accessed 9 October 2019).
[22] "U.S. Mining and Mine Clearance in North Vietnam" (https://www.history.navy.mil/content/history/nhhc/research/library/online-reading-room/title-list-alphabetically/u/u-s-mining-and-mine-clearance-in-north-vietnam.html: accessed 10 October 2019).
[23] Ibid.

CHAPTER 19 NOTES:
[1] USS *Mattaponi* (AO-41) Western Pacific 1968-1969 cruise book.

[2] "Ships of the U.S. Navy, 1940-1945, AO -- Fleet Oilers" (https://www.ibiblio.org/hyperwar/USN/ships/ships-ao.html: accessed 14 October 2019).
[3] "Ships of the U.S. Navy, 1940-1945, AO -- Fleet Oilers"; "Fleet Oilers (AO) Underway Replenishment Oilers (T-AO) Index" (http://www.navsource.org/archives/09/19/19idx.htm: accessed 14 October 2019).
[4] Ut supra.
[5] Thomas Wildenberg, *Gray Steel and Black Oil Fast Tankers and Replenishment at Sea in the U.S. Navy, 1912-1995* (Annapolis, Md: Naval Institute, 1996), 100.
[6] "Jumboized Tankers" (http://navy.memorieshop.com/Story/NFAF.html#Jumboized); "Class: *Ashtabula* (AO-51)" (http://www.shipscribe.com/usnaux/AO/AO51.html: accessed 15 October 2019).
[7] Ut supra.
[8] *Ashtabula*, DANFS.
[9] "*Neosho* Class Naval Fleet Oiler" (http://navy.memorieshop.com/Neosho/Class.html: accessed 13 October 2019).
[10] USS *Ponchatoula* (AO-148) Western Pacific 1971-1972 cruise book.
[11] Ibid.
[12] "1969 Historical Report from the Task Force 130 Manned Spacecraft Recovery Office"; Apollo Recovery Operational Procedures Manual (http://www.collectspace.com/review/apollo-recovery-operational-procedures.pdf); "Hawaii's role in NASA's space exploration programs" (https://www.nasa.gov/feature/hawaii-s-role-in-nasa-s-space-exploration-programs: accessed 17 October 2019).
[13] Ut supra.
[14] Ut supra.

CHAPTER 20 NOTES:
[1] USS *Sacramento* (AOE-1) Western Pacific 1966-1967 cruise book; *Sacramento*, DANFS.
[2] *Sacramento*, DANFS.
[3] USS *Sacramento* (AOE-1) Western Pacific 1966-1967 cruise book.
[4] *Sacramento*, DANFS.
[5] Ibid.
[6] Ibid.
[7] Ibid.
[8] Ibid.
[9] "USS *Camden* (AOE-2)" (http://www.navsource.org/archives/09/59/5902.htm: accessed 24 September 2019).
[10] USS *Camden* (AOE-2) Western Pacific 1970-1971 cruise book.
[11] USS *Camden* (AOE-2) Western Pacific 1970-71 cruise book, and Western Pacific 1974-1975 cruise book.

[12] "Navy Replenishment Oiler *Wichita* (1968)" (https://www.maritime.dot.gov/sites/marad.dot.gov/files/docs/about-us/history/vessels-maritime-administration/911/wichitahaerreport.pdf: accessed 27 September 2019).
[13] Ibid.
[14] USS *Wichita* (AOR-1) Western Pacific 1971-1972 cruise book.
[15] *Wichita, DANFS*.
[16] Ibid.
[17] Ibid.
[18] "Gen. Vo Nguyen Giap, Who Ousted U.S. From Vietnam, Is Dead" by Joseph R. Gregory, *New York Times*, 4 October 2013.
[19] Bruhn, *Wooden Ships and Iron Men, Volume III*, 197; John Darrell Sherwood, *Nixon's Trident: Naval Power in Southeast Asia, 1968–1972* (Washington, DC: Naval History and Heritage Command, 2009), 35.
[20] Sherwood, *Nixon's Trident*, 35.
[21] "Cold War Warriors, An Amazing Day, The final major naval battle of the Vietnam War" by Michael Cuseo (http://www.emmitsburg.net/archive_list/articles/misc/cww/2012/amazing_day.htm: accessed 20 October 2018).
[22] Edward J. Marolda, *By Sea, Air, and Land*, Chapter 4: Winding Down the War, 1968 – 1973.
[23] Ibid.
[24] Ibid.

CHAPTER 21 NOTES:
[1] Bruhn, *Wooden Ships and Iron Men, Volume III*, 195.
[2] Marolda, *By Sea, Air, and Land*, Chapter 4: Winding Down the War, 1968 – 1973; Bruhn, *Wooden Ships and Iron Men, Volume III*, 227.
[3] Marolda, *By Sea, Air, and Land*, Chapter 4: Winding Down the War, 1968 – 1973.
[4] "Vietnam War Peace Talks" (https://alphahistory.com/vietnamwar/vietnam-war-peace-talks/: accessed 23 May 2019)
[5] Bruhn, *Wooden Ships and Iron Men, Volume III*, 197.
[6] Marolda, *By Sea, Air, and Land*, Chapter 4: Winding Down the War, 1968 – 1973.
[7] Bruhn, *Wooden Ships and Iron Men, Volume III*, 197; Sherwood, *Nixon's Trident*, 35; "Cold War Warriors, An Amazing Day, The final major naval battle of the Vietnam War" by Michael Cuseo (http://www.emmitsburg.net/archive_list/articles/misc/cww/2012/amazing_day.htm: accessed 20 October 2018).
[8] Sherwood, *Nixon's Trident*, 45.
[9] Ibid.
[10] "A Chronology of the U.S. Navy in Vietnam and Southeast Asia, 1950–75" (https://www.history.navy.mil/browse-by-topic/wars-conflicts-and-operations/vietnam-war0/chronology.html: accessed 20 October 2018);

Edward J. Marolda, *By Sea, Air, and Land*, Chapter 4: Winding Down the War, 1968 – 1973.

APPENDICES NOTES:
[1] Chief of Naval Operations, Master List of Unit Awards and Campaign Medals, OPNAVNOTE 1650 N09B1, 9 March 2001.
[2] "Com7th Fleet Vietnam Ship Casualties" (http://www.ussmansfield.com/7thfleet/: accessed 18 July 2019).

Index

Abrams Jr., Creighton W., 219
Aldrin, Edwin "Buzz," 89
Anders, William, 82
Anderson, Dick, 188
Angeloff, Ed, 102
Armstrong, Neil, 89
Australia/Australian
  Australian National Line (ANL), 24, 30-31
  Military
    1st Australian Task Force (1 ATF), 26, 33
    Australian Army, 24, 138
      1st Armoured Regiment (1 AR), 138
      Royal Australian Air Force (RAAF), 28, 30, 31
        No. 9 Squadron, 28
        No. 2 Squadron, 30
      Royal Australian Artillery 104 Field Battery, 28
      Royal Australian Navy (RAN)
        Clearance Diving Team
          Three (CDT 3), 28, 133-142
        Helicopter Flight Vietnam, 133
          No. 725 Squadron, 26
        HMAS Lonsdale (naval depot), 32
      Royal Australian Regiment, 24, 134
        1st Battalion (1RAR), 24-25
        4th Battalion (4RAR), 28
        5th Battalion (5RAR), 6th Battalion (6RAR), 26
  New South Wales, Sydney, 24, 34, 142, 259-262
  Queensland, Cairns, Trinity Bay, 30
  South Australia, Wallaroo, 32
  Victoria
    Corio Bay, 30
    Melbourne, 30-32
Avery, Colin, 129
Bakutis, Fred E., 202
Ball, Arthur W., 102, 273
Barth, Bruce, 63, 70
Bavarik, Andy, 103
Bland, Richard, 102-103
Boettcher, Peter, 135
Borman, Frank, 82
Bourke, Richard, 34

Boyd Jr., John Huntly, 121
Brophy, Paul, 48
Brown, Robert B., 150
Bruce (Ensign), 62
Brunot, Richard, 12
Buckner Jr., Simon B., 71
Bundy, McGeorge, 17
Burnett, Patrick Richard, 29-31
Campkin, John, 129
Cass, Eugene, 103
Cernan, Eugene A., 84-85
China
  Hainan Island, 63
  Kiang Yin, Nanking, Rose Island (Leigong Dao), 146
  Shanghai, Tsingtao, 144-148
Churchill, Winston, 11
Clark, Brian V., 135
Collingwood, John F., 150-151
Collins, Michael, 89
Crandal, Carol A., 152
d'Argenlieu, Thierry, 11
Dai, Bao, 11-13
Dau, Heinrich, 35
Digney, Larry J., 135, 139
Draper, Arthur J., 152
Duffner, Gerald J., 151
Dunn, James V., 55
Eden, Barbara, 203-206
Eisenhower, Dwight D., 76
Ey, Anthony L., 135-142, 273
Franklin, Thomas, 103
Furner, Brian John, 135-142
Giap, Vo Nguyen, 11-13, 216
Gilchrist, John Joseph, 135-142
Gile, Chester, 178
Green, Leonard, 168
Grimanes, P. 30
Gryniewicz, Paul, 103, 105
Gummoe, John Claude, 167-168
Hall, George Henry (1st Viscount, PC), 145, 148
Heath, Donald R., 15
Heatherman, John, 101
Hewitt, William, 2, 7
Hood, John O., 135, 137
Holt, Harold, 26
Hong Kong, Kau yi chau Island, 108

Index   305

Hope, Bob, 205-206
Isaman, Roy Maurice, 65
Japan
    Hiroshima, 149
    Nagasaki, 149, 244
    Sasebo, 80, 110-111, 121, 128, 156-157, 169, 212, 227, 248
    Tokyo, 100, 245
    Wakayama, 144, 149
    Yokosuka, 57-60, 77, 80, 100, 127, 145-146, 157, 160, 181, 212, 247
Johnson, Lyndon B., 17, 20-24, 216-220
Kai-shek, Chiang, 11
Kane, Bill, 23
Kember, Philip C., 135-137
Kerans, John Simon (Comdr., DSO, RN, MP), 146
King Jr., Jerome H., 177
Kissinger, Henry, 90, 217-222
Kruder, Ernst-Felix, 35
Kuhn, Rik, 91
Lamberson, Phil, 51, 54
Lapu-Lapu (Chief), 42
Lassau, Geoffrey D., 135
Linton, Edward ("Jake") Wilfred (Comdr., BEM RAN (Retired), 135-142
Lodge, Henry Cabot, 220
Long, Robert L. J., 90
Lovell, James, 82
Lukacs, Mary C., 152
Magellan, Ferdinand, 42
Magnuson, Peter A., 135-137
Mann, Russ, 75
Manwaring, Percival C. W., 35
Martinez, Roland Nino, 74
McCain Jr., John S., 90
McCall, John, 103
McCauley, Brian, 223
McDonald, David L., 19
McGonagle, William L., 184
McNamara, Robert, 17
Menge, Robert Frederick, 152
Minh, Ho Chi, 9-12, 216
Mitchell, Eugene B., 119-120
Moorer, Thomas H., 16, 19
Narramore, Phillip C., 135-142
Nation, William C., 135
Nixon, Richard M., 82-90, 185, 216-222
Norway, Sorgulenjord, 35
O'Daniel, John W., 15

Operation
  BEAU CHARGER, 153
  BEAVER CAGE, 152
  CIMARRON, 155
  DECKHOUSE V, 130
  FLAMING DART, 17
  GAME WARDEN, 2-3, 20, 273
  LINEBACKER, 185-187
  MARKET TIME, 19-20, 41, 43, 63-72, 82, 90, 111, 118, 191, 200, 203, 273
  OLYMPIC, 149
  PASSAGE TO FREEDOM, 14
  ROLLING THUNDER, 18
  SEA DRAGON, 90, 134
Paddock, Gary C., 138
Palmer, Philip M., 185-187
Paul (Mr.), 37
Philip, A. A. C., 34
Philippines
  Baguio City, 46, 188, 256
  Cebu Island, Mactan Island, 42
  Manila, 5, 46, 75, 176, 188
  Olongapo, 188, 253, 255
Puccini, Don, 48
Quinn, Janice M., 152
Reilly, Robert Francis, 38
Risner, Robinson, 189
Rogers, William P., 90
Sabin Jr., Lorenzo S., 15
Salzer, Robert S., 177
Schmitt, Alfons, 35-36
Searle Jr., Willard, 119-120
Shelton, Doniphan B., 190

Ships and Craft
  Australian
    Anzac, 25
    Boonaroo, 29-36, 134
    Brisbane, Perth, 134
    Duchess, Melbourne, Parramatta, Vampire, 24-26
    Hobart, 134
    Jeparit, 26-36, 134
    Supply, 24
    Sydney, 23-36, 134
    Vendetta, 26, 134
  British
    Ajax, Cornwall, Exeter, 35
    Amastra, Helisoma, 129

Amethyst, Black Swan, Consort, London, 145-146
Dutch, *Kara*, 129
German, *Admiral Graf Spee, Altmark, Bismarck, Pinguin*, 35-36
New Zealander, *Achilles*, 35
Panamanian
  Impala, 39
  Sea Raven, 127-128
Soviet, *Gidrofon*, 132
United States (American)
  Army
    Davidson, 132
    YFU-63, 135-140
  Coast Guard, *Point Kennedy*, 51, 54, 55
  Merchant Marine/Military Sea Transportation Service (MSTS)
    General A. W. Brewster, 15
    General William A. Mann, 21
    Saumico, 91
    Kingsport, 80
  Navy
    amphibious
      amphibious assault ships (LPH)
        Guadalcanal, Princeton, 83, 85
        Iwo Jima, 122
        Okinawa, 153, 206
      attack cargo ships (AKA/LKA)
        Montague, 15
        Rankin, 83
      attack transports (APA/LPA)
        Bayfield, Menard, 15
        Chilton, Francis Marion, Sandoval, 83
      tank landing ships (LST)
        Caroline County, Coconino County, Snohomish County, 130-132
        Floyd County, 67
        Garrett County, 2-3, 7
        Harnett County, 2, 54-55
        Hunterdon County, Jennings County, 2
        LST-516, 14
        Mahnomen County, 158-159
        Outagamie County, Summit County, Terrell County, 126-128
      transport (AP), *Geiger*, 131-132
    auxiliaries and service vessels
      ammunition ships (AE)
        Butte, 184-185, 224, 267
        Chara, 176-177, 267, 275
        Diamond Head, 178-180, 268
        Firedrake, 40, 178-179, 268

308   Index

Flint, 184-185, 224, 268
Great Sitkin, 178-179, 268
Haleakala, 182-183, 268
Kilauea, 173, 183-185, 268
Mauna Kea, 181, 185, 223, 269
Mauna Loa, 173, 269, 273
Mazama, 173-175, 269
Mount Baker, 172-173, 269
Mount Hood (AE-11 and AE-29), 171-185, 224, 269
Mount Katmai, 178-185, 223, 269, 273
Nitro, 179-185, 223, 269
Paricutin, 178-179, 269
Pyro, 182, 269
Rainier, 172-179, 269
Santa Barbara, 184-185, 223, 270
Shasta, 172-174, 270
Suribachi, 181-185, 223, 270
Vesuvius, 178-185, 223, 270
Virgo, 4, 175-176, 270
Wrangell, 173-179, 270
auxiliary fleet tugs (ATA)
  Mahopac, 125-127, 268
  Sunnadin, Tillamook, Wandank, 125, 270
combat stores ships (AFS)
  Concord, San Diego, Sylvania, 118
  Mars, 44, 108, 117-118, 269
  Niagara Falls, 108, 117-118, 224, 269
  San Jose, White Plains, 108, 118, 224, 270
destroyer tenders/repair ships (AD, AR)
  Ajax, 158, 164-170, 267, 273
  Hector, 158, 169-170, 268
  Jason, 157-158, 167-170, 268, 273
  Delta, 158, 168-169, 233, 268
  Isle Royale, 158, 268
  Klondike, 158, 168, 268
  Markab, 158-163, 226, 268
  Piedmont, 158, 169-170, 269
  Samuel Gompers, 158, 169-170, 270
environmental research (intelligence-gathering) ships (AGER)
  Banner, Pueblo, 3-5
fast combat support ships (AOE)
  Camden, 207-213, 218, 267
  Sacramento, 39-44, 207-214, 218, 224, 270
fleet oilers (AO)
  Ashtabula, 195-205, 225, 267, 273
  Cacapon, 197, 224, 267

Caliente, 197, 224, 267, 273
   Chemung, 197, 267
   Chipola, 83, 197-203, 267, 273
   Chuckawan, 83, 202-203
   Cimarron, 195-199, 267
   Guadalupe, 197, 268
   Hassayampa, 83, 198-203, 223, 268
   Kawishiwi, 198, 268
   Kennebec, 193-194, 268
   Manatee, 197, 223, 268
   Marias, 197, 224, 268
   Mattaponi, 191-195, 227, 269
   Mispillion, 196-197, 269
   Navasota, 196-197, 269, 273
   Neches, 194-195, 269
   Passumpsic, 196-197, 223, 269
   Platte, 197, 269
   Ponchatoula, 198-203, 224, 269
   Taluga, 197, 270
   Tappahannock, 194-195, 228, 270
   Tolovana, 198, 224, 270
   Waccamaw, 196-198, 223, 270
fleet ocean tugs (ATF)
   Abnaki, 124, 132, 267
   Apache, Arikara, 124, 267
   Chowanoc, 124, 224, 267
   Cocopa, 120-124, 267
   Cree, 205
   Hitchiti, 124-131, 268
   Lipan, 124, 268
   Mataco, Quapaw, 124, 269
   Moctobi, 124, 224, 269
   Molala, 124, 127, 269
   Munsee, 121-124, 269, 273
   Salinan, 83
   Shakori, Sioux, Tawakoni, Tawasa, Ute, 124, 270
gasoline tankers (AOG)
   Elkhorn, Kishwaukee, Noxubee, Patapsco, 91-106, 268-269
   Genesee, 92-93, 100-106, 268, 273
   Tombigbee, 92, 95, 106, 270
general stores-issue ships (AKS)
   Castor, 109-114, 226-257, 267, 273
   Pollux, 109, 269
hospital ships (AH)
   Benevolence, 145
   Repose, 144-156, 270

Sanctuary, 143-156, 233, 256, 270, 275, 273
light cargo ships (AKL)
  Brule, 1-8, 267, 273
  Deal, Estero, Hewell, Ryer, Sharps, 3-4
  Mark, 3-8, 268, 273
major communications relay ships (AGMR)
  Annapolis, 79-81, 90, 229, 267
  Arlington, 80-90, 267
replenishment oilers (AOR)
  Kansas City, Milwaukee, Savannah, 208, 215, 218, 268-270
  Wabash, Wichita, 208, 214-224, 270
salvage ships (ARS)
  Bolster, 123-128, 267
  Conserver, 120-124, 268
  Current, 124, 129, 268
  Deliver, Grasp, Opportune, 124, 268-269
  Grapple, 105, 120-24, 139, 268
  Reclaimer, 124-128, 269
  Safeguard, 124, 224, 270
seaplane tenders (AV)
  Currituck, 64-72, 268
  Pine Island, 64-71, 269
  Salisbury Sound, 64-69, 270, 273
stores ships (AF)
  Aludra, Bellatrix, Graffias, Procvon, Regulus, Zelima, 108, 267-270
  Pictor, 107-117, 225, 269
  Vega, 107-108, 270, 273
submarine rescue ships (ASR)
  Chanticleer, Coucal, Florikan, 125, 267-268
  Greenlet, 120-129, 268
survey ships (AGS)
  Maury, 37-58, 62, 219, 269
  Rehoboth, 38, 50, 270
  Serrano, 37-53, 62, 219, 270
  Sheldrake, 38, 59-60, 270
  Tanner, 38, 59-62, 270
  Towhee, 38, 60, 270
technical research ships (AGTR)
  Georgetown, 75
  Jamestown, 73-78, 268
  Liberty, 184
  Oxford, 73-78, 269
unclassified vessel (IX), *Kailua*, 173
combatants
  aircraft carriers (CV, CVA, CVE)
    America, 179

Index   311

    Constellation, Kitty Hawk, 174, 217
    Coral Sea, 17-18, 174, 185, 217
    Enterprise, 174, 194
    Franklin D. Roosevelt, 174
    Hancock, 17, 207, 213, 217
    Hornet, 83, 89-90
    Midway, 122
    Oriskany, 21, 174, 205
    Ranger, 17, 174
    Saratoga, 217
    Yorktown, 83
  battleships (BB)
    Kentucky, 208, 213
    New Jersey, 148-149, 183, 187
  cruisers (CG, CGN)
    Bainbridge, 194
    Boston, 167
    Canberra, 21
  destroyers (DD, DDG, DE, DER)
    Carpenter, 83, 213
    Cogswell, 123
    Frank Knox, 120-123
    Goldsborough, New, Nicholas, Rich, Rupertus, 83
    Henderson, 40
    Henry W. Tucker, 21-22
    Robison, 207
 mine warfare
  mine countermeasures support ship (MCS), *Ozark*, 83
  minesweepers (MSO, MSC)
    Conquest, Esteem, Gallant, Illusive, Pledge, 40-41
    Woodpecker, 157
  minesweeping boats
    MSB-22, MSB-32, MSB-45, MSB-49, MSB-51, 135-138

Shotter, Michael T. E., 135
Sihanouk, Norodom, 12
Skinner, Bernard Moreland (Lt. Comdr., DSO, RN), 146
Smith, Chuck, 73
Smith, Larry, 106
Spruance, Raymond A., 149
Stockdale, James, 189
Stafford, Thomas P., 84-85
Storres, Aaron P., 15
Sutherland Jr., James W., 143
Taiwan/Taiwanese
  Kaohsiung, 80, 117, 157, 160, 200, 227, 250
  Keelung, 80

Underwater Demolition Team (UDT), 121
Thieu, Nguyen Van, 85-87
Tse-Tung, Mao, 9
Truman, Harry S., 11
Typhoon
  IRMA, 49
  GILDA, HARRIET, 122
  PATSY, 140
  TESS, 203-204
United States
  Army
    1st Infantry Division, 166
    5th Special Forces, 129
    9th Infantry Division, 87
    101st Airborne Division, 50
    173rd Airborne Brigade, 20-21
    Camp Holloway, 17-18
    Camp McDermott, 129
    Sa Huynh camp, 96-100, 105
    Tenth Army, 71
  Marine Corps
    1st Marine Division, 3rd Marine Division, 100
    Third Marine Amphibious Force, 99
    9th Marine Expeditionary Brigade, 38
    Combat Base, Con Thien, 155
  Navy
    Explosive Ordnance Disposal (EOD), 104, 135-142
    Fifth Fleet, 149
    Harbor Clearance Unit One (HCU-1), 120, 127, 159, 271
    Hawaiian Sea Frontier/14th Naval District, 202
    Helicopter
      Anti-Submarine Squadron Four (HS-4), 88
      Attack Squadron (Light) Three (HAL-3), 3, 166-167
      Combat Support Squadron 1 (HC-1), 137
    Manned Space Recovery Force, Pacific (TF 130), 82-83, 199, 202-203
    Military Sea Transportation Service (MSTS)/later renamed Military Sealift Command (MSC), 21, 36, 146
    Mine Division
      73, 41
      112, 166
    Mine Squadron 11 Detachment Alfa, 137-138
    Naval Air Station
      Cubi Point, 188
      Midway, 86
      Sangley Point, 74-75
    Naval Communications Station

Cam Ranh Bay, 90
Honolulu, 85-86
Naval Forces Philippines, 74
Naval Magazine, Subic Bay, 185-190
Naval Support Activity
  Da Nang, 94, 99, 141
  Chu Lai, 159
  Cua Viet, 101
  Dong Ha, 99-100, 130-131
  Nha Be, 137
  Phu Bai/Hue/Tan May, 99
  Saigon, 5, 170, 273
Patrol Force, Seventh Fleet, 64-65
Patrol Squadron
  VP-40, 63, 72
  VP-46, 74
  VP-47, 72
  VP-50, 69
River Patrol Section 523, 3
Seabees (Naval Construction Battalion personnel), 51, 187
Service Group Three, 5, 60-61, 90, 120, 169
Service Squadron
  Five, 93
  Ten, 144
Underwater Demolition Team (UDT) 11, 84-85
Van Doren, Mamie, 61-62
Veth, Kenneth L., 16

Vietnam
 South
  An Thoi, 4, 113, 115
  Baria, 138
  Bien Hoa, 20-12
  Binh Thuy, Cat Lo, Chau Doc, 4
  Cam Ranh Bay, 30-31, 38, 42-47, 50, 56, 63-75, 80, 90, 113, 115, 128
  Cape Padaran, 50
  Cape Varella, 39
  Chu Lai, 46-48, 99-100, 115, 127-128, 147, 152, 158-159, 273
  Con Son Island/Con Dao Archipelago), 45, 66-67
  Cu Lao Cham Island, 66
  Cua Viet, 91-106, 130-131
  Dam An Hai, 50
  Da Nang, 10, 17, 19, 38, 47-53, 63, 66, 91-105, 113, 115, 130-131, 139-141, 147, 152-159, 169
  Dong Tam, 5
  Duc Pho, 100
  Hoi An, 66

Hon Ngoi Island, 43
Hue, 48, 50, 53, 99-100, 139
Khe Sanh, 81, 217
Military
  Republic of Vietnam Air Force (RVNAF)
    Phan Rang Air Base, 30, 43, 50
    Song Cua Dai Junk Base, 53
    Tan Son Nhut Air Base, 16
My Tho, 5, 7
My Thuy, 105
Nha Be, 5, 136-138, 165-166
Nha Trang, 43, 51, 56-57, 126-129
Nui Dat, 26
Phan Thiet, 50
Phu Bai, 99, 152
Phu Quoc Island, 77-78
Quang Nam Province, 94
Quang Ngai Province, 94, 96
Quang Tri Province/City, 94, 217, 273
Qui Nhon, 100, 115, 126
Royal Bishop Banks, 45
Rung Sat Special Zone, 137, 164-166
Sa Dec, 7
Saigon, 5, 10-16, 20, 44, 46, 66, 136, 142, 164-170
Tan My, 100, 105, 139
Thang Hai, 21
Thon, 105
Thua Then Province, 94
Tuy Hoa, 126
Vinh Long, 1, 5
Vung Ro Bay, 19
Vung Tau, 5, 21-34, 44, 46, 53-56, 62, 113, 134, 141-142, 157-169
North
  Cahn Hoa, 17
  Dong Hoi, 17, 217
  Haiphong Harbor, 11, 14, 134, 185-189, 217-223
  Hanoi, 9-11, 104, 185, 189, 216-221
  Lang Dang, Long Dun Kep, Thai Nguyen, Thanh Hoa, 217
  North Vietnamese Army (NVA)
    304th Division, 308th Division, 217
    429th Sapper Group, 104
  Viet Cong, 409th Battalion, 17
  Vinh, 134, 217

Waddell, Russell, 102
Ward, Norvell G., 16, 60-61, 137
West, Raymond T., 69

Westmoreland, William C., 16-17, 22, 165
Wheelock, Dale, 188
Wilson, David, 168
Winter, Robert, 33
Young, John W., 84-85
Young, Thomas L., 6-7
Zumwalt Jr., Elmo R., 78, 177, 273

About the Author

Commander David D. Bruhn, U.S. Navy (Retired) served twenty-two years on active duty and two in the Naval Reserve, as both an enlisted man and as an officer, between 1977 and 2001.

Following completion of basic training, he served as a sonar technician aboard USS *Miller* (FF-1091) and USS *Leftwich* (DD-984). He was commissioned in 1983 following graduation from California State University at Chico. His initial assignment was to USS *Excel* (MSO-439), serving as supply officer, damage control assistant, and chief engineer. He then served in USS *Thach* (FFG-43) as chief engineer and Destroyer Squadron Thirteen as material officer.

After graduation from the Naval Postgraduate School, Commander Bruhn was assigned to Secretary of the Navy and Chief of Naval Operations staffs as a budget analyst and resources planner before attending the Naval War College in 1996, following which he commanded the mine countermeasures ships USS *Gladiator* (MCM-11) and USS *Dextrous* (MCM-13) in the Persian Gulf.

Commander Bruhn's final assignment was executive assistant to a senior (SES 4) government service executive at the Ballistic Missile Defense Organization in Washington, D.C.

Following military service, he was a high school teacher and track coach for ten years, and is now a USA Track & Field official. He lives in northern California with his wife Nancy and has two grown sons, David and Michael.

Heritage Books by Cdr. David D. Bruhn, USN (Retired)

Battle Stars for the "Cactus Navy":
America's Fishing Vessels and Yachts in World War II

Enemy Waters:
Royal Navy, Royal Canadian Navy, Royal Norwegian Navy,
U.S. Navy, and Other Allied Mine Forces Battling the
Germans and Italians in World War II
Cdr. David D. Bruhn, USN (Retired) and Lt. Cdr. Rob Hoole, RN (Retired)

Eyes of the Fleet:
The U.S. Navy's Seaplane Tenders and Patrol Aircraft in World War II

Gators Offshore and Upriver:
The U.S. Navy's Amphibious Ships and Underwater Demolition Teams,
and Royal Australian Navy Clearance Divers in Vietnam

Home Waters:
Royal Navy, Royal Canadian Navy, and U.S. Navy
Mine Forces Battling U-Boats in World War I
Cdr. David D. Bruhn, USN (Retired) and Lt. Cdr. Rob Hoole, RN (Retired)

Ingram's Fourth Fleet:
U.S. and Royal Navy Operations Against German Runners,
Raiders, and Submarines in the South Atlantic in World War II

MacArthur and Halsey's "Pacific Island Hoppers":
The Forgotten Fleet of World War II

Nightraiders:
U.S. Navy, Royal Navy, Royal Australian Navy, and
Royal Netherlands Navy Mine Forces Battling the
Japanese in the Pacific in World War II
Cdr. David D. Bruhn, USN (Retired) and Lt. Cdr. Rob Hoole, RN (Retired)

On the Gunline:
U.S. Navy and Royal Australian Navy Warships off Vietnam, 1965–1973
Cdr. David D. Bruhn, USN (Retired) and
STGCS Richard S. Mathews, USN (Retired)

Support for the Fleet:
U.S. Navy and Royal Australian Navy Service
Force Ships That Served in Vietnam, 1965–1973

We Are Sinking, Send Help!:
The U.S. Navy's Tugs and Salvage Ships in the African,
European, and Mediterranean Theaters in World War II

Wooden Ships and Iron Men:
The U.S. Navy's Ocean Minesweepers, 1953–1994

Wooden Ships and Iron Men:
The U.S. Navy's Coastal and Motor Minesweepers, 1941–1953

Wooden Ships and Iron Men:
The U.S. Navy's Coastal and Inshore Minesweepers,
and the Minecraft that Served in Vietnam, 1953–1976

www.ingramcontent.com/pod-product-compliance
Lightning Source LLC
Chambersburg PA
CBHW071953220426
43662CB00009B/1117